Translational Systems Sciences

Volume 15

In 1956, Kenneth Boulding explained the concept of General Systems Theory as a skeleton of science. He describes that it hopes to develop something like a "spectrum" of theories—a system of systems which may perform the function of a "gestalt" in theoretical construction. Such "gestalts" in special fields have been of great value in directing research towards the gaps which they reveal.

There were, at that time, other important conceptual frameworks and theories, such as cybernetics. Additional theories and applications developed later, including synergetics, cognitive science, complex adaptive systems, and many others. Some focused on principles within specific domains of knowledge and others crossed areas of knowledge and practice, along the spectrum described by Boulding.

Also in 1956, the Society for General Systems Research (now the International Society for the Systems Sciences) was founded. One of the concerns of the founders, even then, was the state of the human condition, and what science could do about it.

The present Translational Systems Sciences book series aims at cultivating a new frontier of systems sciences for contributing to the need for practical applications that benefit people.

The concept of translational research originally comes from medical science for enhancing human health and well-being. Translational medical research is often labeled as "Bench to Bedside." It places emphasis on translating the findings in basic research (at bench) more quickly and efficiently into medical practice (at bedside). At the same time, needs and demands from practice drive the development of new and innovative ideas and concepts. In this tightly coupled process it is essential to remove barriers to multi-disciplinary collaboration.

The present series attempts to bridge and integrate basic research founded in systems concepts, logic, theories and models with systems practices and methodologies, into a process of systems research. Since both bench and bedside involve diverse stakeholder groups, including researchers, practitioners and users, translational systems science works to create common platforms for language to activate the "bench to bedside" cycle.

In order to create a resilient and sustainable society in the twenty-first century, we unquestionably need open social innovation through which we create new social values, and realize them in society by connecting diverse ideas and developing new solutions. We assume three types of social values, namely: (1) values relevant to social infrastructure such as safety, security, and amenity; (2) values created by innovation in business, economics, and management practices; and, (3) values necessary for community sustainability brought about by conflict resolution and consensus building.

The series will first approach these social values from a systems science perspective by drawing on a range of disciplines in trans-disciplinary and cross-cultural ways. They may include social systems theory, sociology, business administration, management information science, organization science, computational mathematical organization theory, economics, evolutionary economics, international political science, jurisprudence, policy science, socioinformation studies, cognitive science, artificial intelligence, complex adaptive systems theory, philosophy of science, and other related disciplines. In addition, this series will promote translational systems science as a means of scientific research that facilitates the translation of findings from basic science to practical applications, and vice versa.

We believe that this book series should advance a new frontier in systems sciences by presenting theoretical and conceptual frameworks, as well as theories for design and application, for twenty-first-century socioeconomic systems in a translational and trans-disciplinary context.

More information about this series at http://www.springer.com/series/11213

Ram Babu Roy • Paul Lillrank
Sreekanth V. K. • Paulus Torkki

Designing Service Machines

Translating Principles of System Science
to Service Design

 Springer

Ram Babu Roy
Rajendra Mishra School of Engineering
Entrepreneurship
Indian Institute of Technology Kharagpur
Kharagpur, West Bengal, India

Sreekanth V. K.
Rajendra Mishra School of Engineering
Entrepreneurship
Indian Institute of Technology Kharagpur
Kharagpur, West Bengal, India

Paul Lillrank
Department of Industrial Engineering and
Management
Aalto University School of Science
Espoo, Finland

Paulus Torkki
Faculty of Medicine, Department of Public
Health
University of Helsinki
Helsinki, Finland

ISSN 2197-8832   ISSN 2197-8840   (electronic)
Translational Systems Sciences
ISBN 978-981-13-0916-8   ISBN 978-981-13-0917-5   (eBook)
https://doi.org/10.1007/978-981-13-0917-5

Library of Congress Control Number: 2018944770

Printed on acid-free paper

This Springer imprint is published by the registered company Springer Nature Singapore Pte Ltd.
The registered company address is: 152 Beach Road, #21-01/04 Gateway East, Singapore 189721, Singapore

*Dedicated to
the budding entrepreneurs involved in
delivering services
to the mankind for improving their quality of
life.*

Foreword

The services sector dominates the global economy and plays an increasingly important role in the economic growth of developing countries like India. It is estimated to contribute around 54.0% of India's gross value added and provide employment to around 28.6% of the total population in 2017–2018. An emphasis is needed to increase employment in services sectors in developing countries. According to a recent report on service sectors by Department of Economic Affairs, Ministry of Finance, Government of India, India's commercial services exports increased from US$ 52.2 billion in 2005 to US$ 161.3 billion in 2016. Various initiatives have been taken by the government to provide an impetus to the services sector including signing comprehensive bilateral trade agreements with several countries. Government of India recognizes the importance of services sectors and provides several incentives to promote growth in a wide variety of sectors such as education, healthcare, information technology, communications, transportation, banking, finance, tourism, engineering, management, and consultancy.

There is a huge opportunity for entrepreneurship in services, especially in healthcare. The global medical value travel (MVT) market is expected to cross US$ 100 billion in 2019, growing at a compounded annual growth rate (CAGR) of 19.4%. India's share was 3.8% of the global medical tourists and 5.5% of the global revenue in 2014. The government has initiated many policies to promote India as a medical value travel destination. However, healthcare is a troubled industry everywhere. Regardless the system and spending level, there are no cases where all constituents, patients, professionals, payers, and politicians would be satisfied. In the developing world, the problems are lack of access and undercare, and the corresponding challenge is to build systems that can deliver affordable care to the masses. In the developed world, the problems are medicalization and overcare, and lifestyle-related diseases lead to unsustainable cost expansion. The challenge is to contain costs while providing individualized care, i.e., mass produced personalized services.

It has become increasingly clear that the traditional ways of designing health finance, service production, and regulatory oversight are not sufficient. Innovation in healthcare is an increasingly urgent task all over the world. Several start-ups are working on addressing the challenges in healthcare service segment in India and abroad. Several initiatives are in place to provide telemedicine and telehealth services, mobile eye clinics by Sankara Eye Hospital, home care services, etc. Mohalla (community) clinics in New Delhi provide free health services to reduce the financial burden on low-income households by saving travel costs and lost wages at a cost to the government of just over INR 100 per patient. However, there are challenges of poor adoption and utilization of healthcare services by the users due to traditional beliefs, lack of trust, resistance to change, and fear of catastrophic failure.

This book looks deeper into healthcare industry which has been overlooked for long by applying knowledge accumulated in business administration, service engineering, and systems sciences in healthcare delivery. Health service production faces the same basic problems of process design and control as any industry. However, a hospital is not a factory. The product is an offer to help in problems of health. The therapies do not have a scientific foundation with the precision of physics. In health service, demand does not behave as in conventional commodity markets; patients may want what they do not need and need what they do not want. Methods that have a proven track record in industry need to be translated to fit healthcare. This is the objective of Translational Systems Science.

The Service Machine is a conceptual tool to help design service production system. The 'machine' metaphor highlights that services need structures and stability in order to adjust to individual cases while reliably repeating best practices. The authors use the bicycle as an illustrative example of a Service Machine. A bicycle is a structure that utilizes basic mechanical principles by arranging key components on a frame to amplify speed for enhanced reach. The bicycle neatly illustrates basic management dilemmas. The device alone cannot create value. The machine must be deployed and put in motion by a user who has taken the trouble learning how to ride. The rider produces both energy and control while having to obey rules determined by the machine, balance, and pedaling motions. A Service Machine alone is not enough. There must be infrastructure, roads, and traffic rules.

The contributors are covering a new concept and new ground by out-of-box thinking and very timely discussion on what can be done in India and beyond no matter where people live to provide valued healthcare at optimum cost, an interesting read. Many services have evolved evolutionary through trial and error. In India it is called 'jugaad', an innovative fix to make things work, or to create something with meager resources. With increasing size complexity increases, and service systems tend to spin out of control. Innovation needs to be accompanied by frugal engineering, a clear focus on what creates value. The Service Machine is thereby a way to both introduce engineering to services and services to engineers.

Managing Director   Surgeon R Admiral V. K. Singh
InnovatioCuris
New Delhi, India

Adjunct Professor – World Health Innovation Network
University of Windsor
Windsor, Canada

Honorary Professor – Australian Institute of Health Innovation
Macquarie University
North Ryde, NSW, Australia

Preface

We started our journey exploring the concepts of service science for our research on designing various healthcare services, e.g., preventive care, maternity care, emergency medical services (EMS), and health insurance for rural India. We were trying to understand the difficulties faced by various stakeholders in the domain of healthcare services and service industry in general. What makes services different from manufacturing? Why do healthcare services lag behind several other service sectors? The deliberations led to recognizing the difficulty faced in understanding the complexities of service systems in general and those that deliver healthcare services in particular. We decided to develop a top-down approach and explore the system to decipher its key components. Further, we wanted to create a methodology to analyze their structural linkages and interfaces of these components of the service system in detail to understand their dynamics in service delivery.

Though manufacturing industry has developed several methodologies to analyze its production system, there is a lack of a proper framework to examine a service production system that leads to observed functionality and customizability. We wanted to develop an analogous methodology to analyze a service production system. We started looking for the similarities and differences between the manufacturing and service production systems. Observing some similarities among services in different domains, we felt a need to develop models to describe these similarities across different types of services. The work towards this book began with a search for a framework to guide our thought process for analyzing a service system and identify opportunities for improvement. Taking cues from manufacturing domain, we focused our attention on conceptualizing a service machine that can perform individual routines and can work together in sync, when integrated, to deliver the intended service to customer satisfaction.

We had extensive deliberations on how to structure this new paradigm in a holistic manner and relate it to the extant literature on the evolution of service production systems. Finally, we zeroed into a structured template to express our conceptual thoughts on the service machine. We discussed a few conceptual tools from literature to build a foundation for deliberating the concept of service machine. We have developed the concept of service machine using these basic conceptual tools and demonstrated its applications in two case studies from healthcare services, namely, emergency medical services (EMS) and emergency department (ED). Finally, we conclude this book with a brief discussion on challenges faced in defining system boundaries and the interface of different subsystems while integrating them into a larger service production system. We hope that the framework developed in this book will help the readers in analyzing existing service systems, setting up service design goals for new services, developing the service blueprint, and strategies to foster collaboration with their partners and create service innovations.

Kharagpur, West Bengal, India   Ram Babu Roy
Espoo, Finland   Paul Lillrank
Kharagpur, West Bengal, India   Sreekanth V. K.
Helsinki, Finland   Paulus Torkki

Acknowledgments

The authors would like to express sincere thanks to the Rajendra Mishra School of Engineering Entrepreneurship (RMSoEE); Indian Institute of Technology Kharagpur, Aalto University, Finland; and University of Helsinki for providing infrastructure and logistics support to facilitate brainstorming and developing the manuscript of this book. We also thank the research students Amrita; Sharad; Zuhair; Saurabh; Surgeon Rear Admiral V.K. Singh, VSM (Retd.); and Dr. Indranath Banerjee who participated in various discussions on understanding the complexity of healthcare services in rural India. The discussion helped in raising several pertinent questions that need the attention of research community to come up with tools that enable entrepreneurs to design service systems for delivering services to the target customers. We extend our heartfelt thanks to Prof. Partha Pratim Das, Head, RMSoEE and Professor, Computer Science and Engineering for the encouragement and continuous support towards generating new ideas for product and service innovations to solve real-life problems. We also wish to thank Dr. Devendra Mishra, Dr. Amarendra Mishra, and Prof. Farrokh Mistree for inspiring us to design tools to help entrepreneurs in creating new businesses.

We express gratitude to all the individuals who shared with us their experiences and insights on service system design at various forums. We also express our sincere thanks to the editors Yutaka Hirachi and Selvakumar Rajendran for their patience and accommodating our requests for extending the deadlines due to delay in developing the manuscript.

We would like to express our thanks to our family members for bearing with us when we had to be away from them working on the manuscript.

Contents

Abbreviations

AI   Artificial Intelligence
ATM   Automated Teller Machine
BMC   Business Model Canvas
CT   Cycle time
DSO   Demand Supply-based Operating
ED   Emergency Department
EMS   Emergency Medical Services
EMT   Emergency Medical Technician
GDL   Goods Dominant Logic
GPS   Global Positioning System
IHIP   Immaterial, heterogeneous, inseparable, and perishable
IoT   Internet of Things
IoE   Internet of Everything
IT   Information Technology
KPI   key performance indicator
NHM   National Health Mission
OECD   Organisation for Economic Co-operation and Development
PIP   Patient-in-process
PROM   Patient-reported outcome measures
RIM   Resource Integration Model
ROI   Return on Investment
SDL   Service Dominant Logic
SM   Service Machine
TH   Throughput
WIP   Work in Progress

Abbreviations

Chapter 1
Introduction

Abstract The concept of service system and service science has evolved over last two decades to counter the Baumol's disease. Productivity increase in service is possible in some cases, but it does not necessarily follow the manufacturing model. Services have been subject to scientific inquiry for generating new knowledge about services, how they can be measured and evaluated, and understanding how they work. Traditional service science literature from marketing perspective assumes that if it could be found out what customers want and are willing to pay for, producers' will scramble to deliver. However, as services grow in complexity, it can't be assumed that merely knowing what should be done would get it done. A systematic exploitation of knowledge is required to design the service production systems. This chapter outlines the need for a new metaphor called Service Machine, intended to work as a quick-and-dirty shorthand tool for aiding production engineering effort towards design of service production systems.

Keywords Service science · Service system · Service machine · Service productivity

For half a century, the service sector has been ill with Baumol's disease. This ailment, made famous by William Baumol in his 1967 (Baumol, 1967) article, states that productivity in services can't improve significantly. They are constrained by the need for personal encounters (dentist), local markets (hairdressers), and the simultaneity of production and consumption (music performance). The classical drivers of productivity, the division of labor, specialization, standardization, and economies of scale can't work their magic. Meanwhile, productivity improvement in manufacturing leads to higher wages. The service sector has to follow the trend to attract labor. The price of services in relation to manufactured goods increases, restricting demand. As the service sector, riddled by Baumol's disease, approaches three-quarters of the volume of economic activities, the future of growth looks indeed bleak.

Baumol's disease is still rampant but not incurable. A report by the Organisation for Economic Co-operation and Development (OECD) (McGowan, Andrews, & Millot, 2017) noted that productivity in leading service companies had increased faster than in manufacturing. In both sectors, the difference between the frontier firms and the laggards is widening, more so in services. Productivity increase in service is possible, but it does not happen across the sector, and not necessarily following the manufacturing model. Further, the advancement of technology has a heterogeneous impact on productivity improvement of different stages in the value chain. Consider the music industry. Orchestras can't improve their time-measured productivity very much. But music can be recorded and then distributed in several ways. In some services trade and distribution matter more than production efficiency.

The possible cure for Baumol's disease is good news for national economies; improvement is possible. To incumbent service providers it is a threat; if you do not improve, somebody else might. Consequently, services have been subject to scientific inquiry with the purpose of generating knowledge of what services are, how they can be measured and evaluated, and how they work. The scientific enquiry led to a concept broadly known as service systems which are defined as dynamic configurations of resources (people, technology, organizations, and shared information) for value co-creation (Maglio & Spohrer, 2008).

Service science, the academic study of services and service systems has long been dominated by marketing. Marketing literature differentiates between customers as a person who pays for the product or services and consumers as those who use the product and services. In this book, we are going to use the term customer to mean both customer and the consumer depending on the context unless stated explicitly. The assumption has been that if it could be found out what customers want and are willing to pay for, producers' will scramble to deliver. As services grow in complexity, it can't be assumed that merely knowing what should be done would get it done. The design of service production systems, Service Operations Management, is a serious science-based production engineering effort. It requires judicious use of available resources to create value at desired place and time. At the same time, the value should be perceived by the customers to improve customer satisfaction.

Most of the large service industries, such as Healthcare, Banking, and Finance, Travel and Tourism, have been around for a while. They have developed from humble origins using trial-and-error and reacting to environmental influences. Many service systems have not been engineered from the root up but evolved through an evolutionary process. New service concepts, such as Facebook, Uber, and Airbnb are celebrated because they are rare. Most services have legacies. Service system design can seldom start from a blank slate. Existing systems need to be improved and redesigned.

To this effect, service engineers need conceptual maps, systematic descriptions of generic service elements that can be scaled down to specific cases.

> Examples of useful conceptual maps are the periodic tables in chemistry, the Linnaean system in biology, double-entry bookkeeping with a profit-and-loss statement and balance sheet, the Balanced Scorecard, and the Business Model Canvas.

The purpose of this book is to develop a conceptual map for the analysis of existing service production systems and the development of improved service production systems. To this end, we use the metaphor Service Machine.

Metaphor is a figure of speech, which makes a comparison between two things that are unrelated but share some common characteristics. A new phenomenon can be made comprehensible by comparing it to something that is already known. The first automobiles were called 'horseless carts'; early mobile phones were 'cordless'. A service production unit can be compared to a machine, as they share some common characteristics, such as input-output interfaces, energy source, a structure and components, memories and storages, and a control unit.

The Service Machine is a technology. Broadly defined, technology comprises the rules and methods of purposeful human activity that are based on the systematic exploitation of knowledge. Knowledge, in turn, is the product of the scientific method; the combination of data and reasoning as empirical observations are structured and interpreted by logic. The basic form of knowledge is 'if A then B with the probability p.' Technologies build on theories about various phenomena. Knowledge about the natural world leads to engineering technologies, what is conventionally used as the narrow meaning of 'technology': devices, new chemical compounds, software and the like. Knowledge about the biological world leads to bio-, such as genetically modified plants, and in the case of the human body, to clinical medicine. Knowledge about the human mind and social systems is used to construct organizations. However, human behavior can't be known and predicted as precisely as that of nature. Services are to a large extent human interaction. Behavioral technologies do not have the accuracy and predictability of physical technologies. Therefore, Service Machine is fuzzy and stochastic.

> The phenomena of the social world of human beings are not known, or perhaps even knowable, with the precision of physics. Nevertheless, some regularities can be understood and exploited. Both casual observation and masses of economic research have shown that people react to their environments and the choices they have. Most people often tend to do what is beneficial to them and avoid things that harm. This phenomenon is called incentives. Incentives have many attributes and complex contingencies that can be classified, measured and shown to have effects on behavior. These effects, contrary to physical mechanisms, are not deterministic. Everybody in the same situation will not react precisely in the same way to similar incentives. People can to a variable degree be ignorant, stupid, or have complex motives. But a stochastic relation can often be established: in situation A most people most of the time do B. Human Resource and executive compensation experts exploit this phenomenon by designing incentive systems, pay packages, bonuses, options, and perks. These are behavioral technologies.

A prepared mind sees what the casual observer ignores. The Service Machine model is intended to work as a quick-and-dirty shorthand for scholars and analysts who want to grasp the essentials on a service production system. In economic history and management, such templates have proven to be useful. Double-entry bookkeeping, the profit and loss statement and the balance sheet are finely structured instruments by which an observer can quickly grasp the economic situation of an enterprise. The Balanced Scorecard adds essential details such as personnel and customer satisfaction. Like these constructs, the Service Machine model gives a structure, prompts an observer to ask questions, fill the boxes with situational data, connect answers, and get a grasp on what a service production system is and how it works.

This book is intended to be used by researchers, practitioners, and entrepreneurs interested in understanding the challenges in designing and managing service systems. While the benefit and the processes are created based on the domain knowledge of each respective field of application, service production systems share some common properties across disciplines. The book provides a general conceptual framework to translate principles of system science and engineering to service design. The framework will be useful for the analysis of an existing service production system and getting insights into the potential scope of improvements in the system. Besides, it would help in recognizing opportunities for value creation and new business model development in the service industry. The primary focus is on the part of the service system that can reproduce the processes involved in service production, called here a Service Machine. Entrepreneurs may find it as a useful tool to examine various ideas while designing new service systems to exploit various entrepreneurial opportunities.

In this book, we first present briefly some generic thinking tools that have been used in constructing the Service Machine model. Then we present the Service Machine template in some detail, and finally apply it to the analysis of a rather complex service system, Emergency Medical Care.

Chapter 2
The Conceptual Tools

Abstract A set of generic thinking tools is required to understand and develop a concept. New concepts are always built upon existing knowledge bases. In this chapter, we discuss various generic thinking tools used for designing and developing the concept of Service Machine. As a service production system designer, one should understand various definitions of services, the concepts of systems and systems thinking to visualize different abstraction levels of a system from micro to macro, and relevant emerging technologies. Technology means a systematic and purposeful attempt to accomplish something based on knowledge about the underlying phenomena, be they natural, biological, or social. Knowledge has three aspects: ontology, epistemology, and dynamics. With knowledge, technologies can be developed. Service machine is proposed as a conceptual tool for understanding the way service production system is organized in a business. A service machine should be designed using socio-technical approach in order to achieve higher productivity and satisfaction of stakeholders. We also discuss how the service machines are related to the existing concept of Business Model Canvas (BMC). This would be helpful in improving existing services and creating new service businesses.

Keywords Scientific method · Systems thinking · Socio-technical approach · Technology development · Business model

2.1 Scientific Thinking

The lasting value of science does not lie in the results. Science is not about finding and proclaiming the truth, but having a method by which the truth can be approached with statements that are increasingly truth-like. Some theories have been with us for a long time, some others come and go. The core is the scientific method, the rules governing observation, data gathering, analysis, and reasoning.

Inductive reasoning starts from observations (data), looks for patterns, and makes inferences about regularities. Deductive reasoning starts with some first principles or accepted conceptualizations, seeks to find their representations in the world, derives hypotheses, and tests them. Abductive reasoning combines the two into an

© Springer Nature Singapore Pte Ltd. 2019

R. B. Roy et al., *Designing Service Machines*, Translational Systems Sciences 15, https://doi.org/10.1007/978-981-13-0917-5_2

interaction between concepts and empirical findings, also known as Grounded Theory (Glaser & Strauss, 1999).

The Service Machine crystallizes research and experience of service production. Thereby it can serve as a conceptual starting point for abductive reasoning.

2.2   Knowledge as a Basis for Technology

Technology means a systematic and purposeful attempt to accomplish something based on knowledge about the underlying phenomena, be they natural, biological, or social. Knowledge has three aspects. Ontology (what is it?), epistemology (what can be known about it?), and dynamics (how does it work?). When you know what something is, how much there is of it, and how it works, you have knowledge about it. With knowledge, technologies can be developed. Technology development is the ultimate purpose of the pursuit of science.

Scientific inquiry typically proceeds in five steps.

First the studied system, the empirical reality in its context, is defined and delineated as the unit of analysis. It is like a playing field with defined boundaries that separate what is included and excluded in the research.

> In the case presented later in this book, the empirical reality is Emergency Medical Services (EMS), i.e., ambulance services, and the Emergency Departments (ED) where patients are taken. It is constructed to respond to urgent and severe medical needs of the patient. The system consists of vehicles (ambulances) and crews, communication devices, a dispatch center, and the ED.

Second, within the unit of analysis, the relevant phenomena are described qualitatively (ontology) by explaining, what they are, which are the components and how they are connected. This produces a descriptive qualitative (conceptual) model.

> Since this is a study in Healthcare Operations Management, the relevant phenomena within the playing field are related to the performance of the EMS/ED system, such as response time, throughput volume (transported and treated patients), and quality.

Third, the unit of analysis is described quantitatively. Relevant phenomena are operationalized as variables, such as how much, how long, how often. This produces a descriptive quantitative model that gives the details about relative weights and volumes of the system components.

> Response time is measured as the time from the first contact to action. Throughput is the number of dispatches and transported patients per day or week, for the ED it is treated, patients. Quality can be measured as deviations from targets, such as mortality and delays; and as the relationship between expectations and experience.

Fourth, the relations between the parts are examined to find out which affects what in which way. The relations can be causal (from every A always follows B, A is a necessary and sufficient explanation of B), stochastic (if A increases there is a high probability that B increases, too), or enabling (A makes B possible but does not guarantee it). This produces a dynamic, or explanatory model telling how the system works.

> In Operations Management there are well-established dynamic relations. One such is Little's Law. It states the relations between work-in-process (WIP), throughput (TH), and cycle time (CT) as WIP = TH × CT. This formula can be rearranged as CT = WIP/TH, and as TH = WIP/CT.

Fifth, a predictive model is based on the dynamic model and allows estimates of what happens if some of the system components are altered.

> Little's law says that if WIP is reduced with constant CT, throughput will decrease.

Finally, prescriptive model builds on predictive models and stipulate, if you do this, then that will happen. A prescriptive model is the prototype of a technology.

> Lean production, also known as the Toyota Production System, originated from the strategic imperative of reducing work-in-process inventories (WIP), to minimize the need for and cost of operating capital embedded in inventories. To make this possible with constant throughput, the thing to do is to shorten cycle times. The reduction of cycle times can be done in many ways, such as reducing time-consuming quality problems through standardization, and by lowering setup-times for machinery.

2.3 Systems and System Thinking

System is both a way to think – to be systematic – and a focus of inquiry. A basic definition of a system includes:

– A system has boundaries. It must be possible to determine which entity is a part of the system and which is not.
– There are two or more distinct elements or components within the boundary.
– Some interactions between the components make it more than the static sum of its parts.
– A system connects to other systems through interfaces.
– Systems can have layers and hierarchies; there may be systems within systems.

An essential aspect of system thinking is the idea of different abstraction levels of a system from micro to macro. A leaf is a system that is part of a tree, which is a part of a forest, which is a part of a landscape. The human body is a system that consists of several organ systems, such as the neurological, cardiovascular, and the immune systems. A task is a part of a process, which is part of a production unit, which is part of a supply network. A Service Machine is a part of a business and eventually a business ecosystem.

System levels are units of analysis. Reductionism is the idea that every complex phenomenon can be adequately explained by analyzing its simplest, most basic mechanisms. Its opposite is holism, the claim that parts are all interconnected and cannot be understood without understanding the entire whole. Reductionism and holism are extreme positions. The middle ground takes more effort than extreme positions. It is to dissect systems into components, understand what they are and how they work, then put them together again as a synthesis, and then connect to the next higher system level.

System layers build on each other's. The interactions between various layers make it difficult to predict the behavior of a whole system merely with the knowledge of the behavior of its individual components. The physical world is a system. Biology and living organisms are systems that can't violate the laws of physics, but neither can they be fully described by physics. Human behavior is founded on biology, but can't be reduced to biological reactions. Societies are made out of human behavior, but not explainable by individual psychology. Each system layer brings specific phenomena and their dynamics.

A particular instance or situation can simultaneously be affected by several system layers. In business there typically are social, technical and economic considerations, each which builds on a specific logic. The social system defines a person's role in various settings, as a family member, employee, citizen, customer, or passenger. Technology both enables and restricts. Economics is present when choices have to be made under scarcity; the cake can't be both kept and consumed. Many problems (waiting time) appear at the individual level, while their causes are at a higher system level (resource allocations), and therefore can be solved only at that level.

> A patient meets a doctor. The social system influences the fears, and hopes of the patient. Both parties have expectations of a proper patient-doctor role-relation set by the given business environmental context. The technological system is knowledge acquired through education, and skills developed through training. There are protocols, codified best practices, devices, and pharmaceuticals. The diagnostic technologies define what can be known about the patient's medical condition, and clinical technologies define what can be done. The economic system stipulates that the doctor expects to be paid, as do all those who have delivered equipment and supplies. An insurer or a government has established rules about costing and pricing. However, everything that is socially preferable and technically possible is not economically affordable. Therefore, the sustainability of such businesses becomes questionable.

2.4   Socio-technical Systems

Existing service design literature puts more emphasis on the design of technical systems (e.g., the design of service concepts, processes, and interfaces), but ignores the design of social systems (e.g., fostering social relationships to achieve the design goals) (Baek, Kim, Pahk, & Manzini, 2017).

Attention must be given to building social networks to facilitate collaboration for service production. A socio-technical view of systems emphasizes that the optimal organizational performance can be achieved by jointly optimizing both the social and technical systems used in a production system (Loudon & Loudon, 2015). A service production system requires interaction between human and technology making it a perfect example of a socio-technical system. A socio-technical view of the system considers both technical and social features of a real-world system and critically evaluates various alternative solutions. It looks for solutions that provide the best fit between the technical system and the social system. The socio-technical design blends technical design efficiency with sensitivity to organizational and human needs of the social design to achieve higher job satisfaction and productivity.

A service production system requires interaction between living and non-living entities and hence behaves as a complex adaptive system. Though the entities involved work towards achieving a defined objective, the outcome of the system is often difficult to predict. For example, a healthcare system has its primary objective to deliver high-quality and cost-effective care to patients. But we do not have enough control over the outcomes of healthcare services. So far we are unable to achieve satisfactory healthcare services across the globe. The roles of different stakeholders, e.g., physician, pharmaceutical companies, hospitals, diagnostics services must be defined precisely to enable them to participate in the creation of health value for the patients. The role of patients and their family also becomes crucial to enable them to participate in the co-creation of health value. The uniqueness of Healthcare service system can be understood by the fact that every patient is unique and need personalized care. All the medical professionals are also unique in terms of their knowledge, skill sets, and style. The uniqueness of patients and professionals leads to a challenging matchmaking problem. Can we match the core competency and unique style of medical professionals to specific needs of patients at an affordable price? Can we use service operations management principles to improve the overall efficiency of the existing system? There is a need for developing a framework to engage medical professionals to the tasks they are best at doing. Service machine should have provision to do appropriate matchmaking under given scenario.

2.5   Information System for Services

Information system plays a crucial role in the effective delivery of any services. It helps in dissemination necessary information to concerned stakeholders to facilitate service co-creation. It also helps in computation and monitoring of various parameters, e.g., key performance Indicators (KPI) that eventually are required to develop benchmarks. For example, recent interest in patient-centric approach strives to drive the industry towards value-based healthcare. Its implementation requires identification of outcome measures that matter to patients really and definition of global outcome standards. A dedicated effort is required to drive adoption of the measures and reporting them worldwide. Judicious use of information system in services entails strategic planning for the selection and implementation of appropriate information technology, organizational structure, and human resources. Top management must ensure that everyone involved in the service production system including the customers receive sufficient training to use the technology adequately.

If we consider the healthcare system that faces difficulty in managing the triangle of quality, cost, and access, the information system can help to bridge the gap. Telehealth solutions being developed and deployed in some countries are bringing down the cost of treatment and increasing the access to quality healthcare. For designing healthcare information system in any settings, we must define the nature and scope of information systems and implement appropriate systems to meet the information needs of stakeholders. Several sub-systems of the healthcare information system can be planned such as video conferencing and live chat options, computerized decision support-support and order entry systems for physicians, electronic health or medical records can provide flexibility to physicians to log in any time and monitor patients. The information system may have tools for patients to access service providers of their choice, to manage their medical information, for seamless billing process, and receive feedback and results instantly.

2.6   Internet of Things (IoT) and Analytics

Internet of things and analytics are emerging technologies having applications in various domains including service industry. It is helping in understanding the customer requirements better and delivering value more effectively. Healthcare information systems are installing, replacing, or updating basic transaction systems – data on patients, care provision processes, and costs creating an Electronic Health Record. Information system also integrates billing and other operational management systems creating a pool of real-life data. There is a need for understanding what these data mean and how analytics on them can translate efforts into better patient care. There are different types of healthcare data and their access: patient portals, mobile patient access to records, telemedicine, etc. There is a requirement for recommendation system for personalized care. However, all the stakeholders speak different languages and see the world from different perspectives. The overall healthcare system behaves as a complex adaptive system and predicting the performance of the whole system becomes difficult. According to the management guru Peter Drucker, healthcare is the most difficult, chaotic, and complex industry to manage and the most complex human organization ever devised.

The healthcare business landscape is changing fast with an increasingly higher mobile population with higher mobile/IT penetration. It requires addressing various needs for location-based services (whether the care is available in the neighborhoods) including addressing "ambulatory" and "inpatient" care needs. There is a time lag between technological innovation and its adoption among end users. It needs insights for creating business innovation that can transform healthcare. There is a need for exchange of knowledge between business and medicine to foster entrepreneurship in healthcare with the availability of financial support to engage in value creation. Healthcare is far behind other industries in terms of analytical sophistication and adoption of analytics in every aspect and segment of the industry. Further, there are silos across clinical, operational, and business groups within healthcare organizations. It provides an excellent opportunity for leveraging analytics to gain insights from big data (challenges pertaining to volume, velocity, variety, and veracity) and the Internet of Things (IoT) based systems for development and analysis of healthcare business model.

Information systems and analytics can help in integrating clinical and business data to assess or compare the cost-effectiveness of clinical interventions or processes. We can monitor key performance metrics across providers, groups, care venues, and locations and identify ways to improve outcomes and reduce unnecessary costs. Analytics can help service providers to reach out to patients to guide them through the treatment processes. Patients can receive the right care, at the right venue, and at the right time with privacy protection. It can help in predicting emergency room visits or heart attacks, predicting risk in major surgery. It can also help in benchmarking hospital services and ranking the safest hospitals as well as profiling surgeons by their complication rates. We may perform an analysis of the distribution of disease and health outcomes in relevant populations of interest (e.g., general population, health system members, patient sub-groups). Clinical analytics is being used in various contexts of quality improvement (e.g., chronic disease, patient utilization, population health, public health). There is a growing interest in development and implementation of informatics and analytics tools for use in healthcare quality and performance improvement.

2.7 Sustainability

Service production system being a the socio-economic-technical system needs to be sustainable – which should balance the triangle of People (social), Planet (environment), and Profit (economy) (PPP). It is important to analyze these dimensions of the system is crucial. As a collaborative value creation process, the society is involved in the service production system. The economic aspects of the service production are complicated. The primary user may not be the payer of the service, but the payer also will be a benefactor. Providing the services for free may not seem to be sustainable, but by designing and developing appropriate revenue model, it would be possible. Google, being as search service machine provide the services to the searchers without collecting money from them. They receive money from advertisers based on the data obtained from the users. Although it seems economically cheaper, it indirectly costs the privacy of the users (the people). It requires enormously large data servers to store the data, and there are chances of contributing to the global warming. Businesses involved in the services will look for their return on investment (ROI). Thus balancing the three aspects of sustainability is not easy, but it is necessary.

2.8　Segmentation

Segmentation is an ontological exercise to divide an entity or a system into smaller chunks for characterization and better understanding. It divides an entity into sub-groups, segments, according to some criteria. In customer marketing, common segmentation principles are age, sex, residence, ethnic group, profession, or income level. A customer segment is in some relevant characteristic more homogeneous than the whole customer population. Thereby variability of demand can be managed by offering segment-specific products. In services, segmentation requires different approaches, as customers participate in the production.

> Health services are segmented following the elements of clinical medicine. There are classifications following organs and organ systems (cardiology, neurology, ophthalmology, etc.), primary causes (trauma, infections, tumors, etc.), principal methods of treatment (surgery, internal medicine, physiotherapy, etc.), and physique present in demographic groups (women, children, elderly, etc.). Overlapping segmentation principles create narrow specialties (neurological trauma surgery for children) that are difficult to integrate into patient episodes.

A comprehensive segmentation method is based on jobs-to-be-done, that is, what needs to be done and what can be done. The entity to be segmented is the constellation of demand and supply, hence the term Demand-Supply -based Operating (DSO) types or logic (Lillrank, 2018).

In Healthcare the demand side is classified along urgency (urgent – can wait), severity (severe – mild) and arrival (random – planned). The supply-side entities are diagnostic clarity (cause and cure known – unknown), endpoint (expected positive – chronic and terminal), risk (realized – elevated), and production constraints (job shop – disconnected flow – connected flow – supply network). The demand and supply side classification factors produce the seven demand supply based operating modes, called DSO -types: prevention, emergency, one-visit, electives, cure, care, and projects as shown in Fig. 2.1.

Emergency services are designed to deal with all types of medical problems that are urgent and perhaps severe. Within this, services never the less need to apply segmentation in terms of the degree of urgency and severity. From this flows triage classifications, expressed on a scale from A (immediate resuscitation) to E (no need for emergency service).

With Emergency Departments incoming patients are typically segmented into pregnancies, wheel-in (ambulances), walk-in, and children, sometimes with separate entrances for each segment. Another principle, applied at the triage, is to separate the cases where the clinical picture is clear (wounds, fractures) from those that require diagnostic data collection (blood analysis, imaging), and clinical decisions.

Service production systems are in many ways different from manufacturing and require different segmentation methods. Segmentation is important for defining the services to be delivered and designing the building blocks of the service machine. It needs further probing to understand the implications of various bases of segmentation on the complexity of service machine design and the overall cost of service delivery.

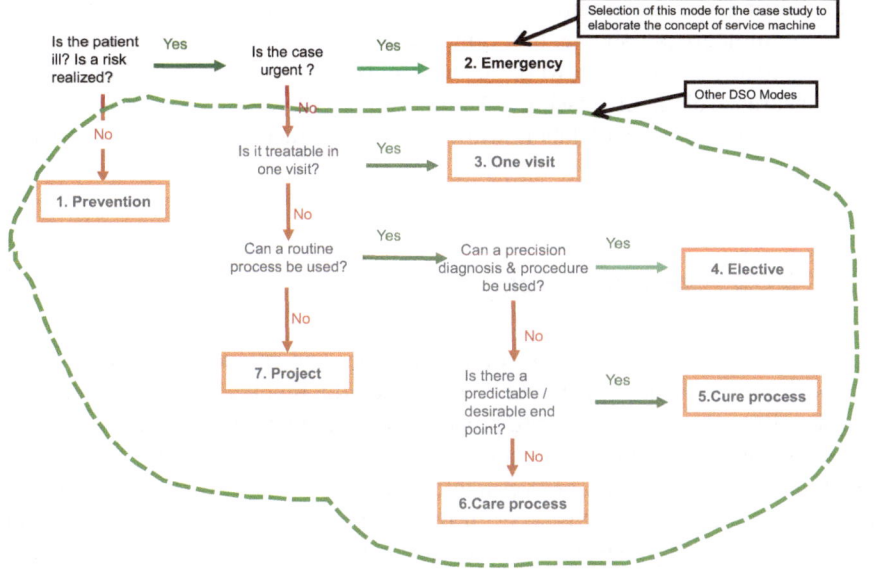

Fig. 2.1 DSO –types. (Adapted from Lillrank, Groop, & Malmström, Milbank Quarterly, 2010)

2.9 Definition of Services

Services as a professional and commercial enterprise are rather recent. The early classical economists were focused on agriculture and manufacturing. Services do not have permanence and will not circulate in the economy as goods do. In those days, the majority of service workers were domestic servants. The employer-employee –relation was embedded in tradition and authority so what would be the need for scientific service management?

With increasing wealth, the service sector expanded and could not be ignored. A novel idea was that immaterial value could be created and traded. Services can't be owned, stolen or returned, but they can be priced, sold, and bought. This line of thinking has been crystallized in the IHIP –definition: services differ from physical products in that they are immaterial, heterogeneous, inseparable, and perishable (Fitzsimmons & Fitzsimmons, 2010; Zeithaml, Bitner, & Gremler, 2012).

2.9.1 Services as Immaterial Products

The IHIP is focused on services like different kinds of products. As there is a broad spectrum of services, from housecleaning to legal advice, from taxis to heart transplants, the definition ran into conceptual confusion and fuzzy borders. A dentist provides a service, but the pain of the drill is very much material. An automatic teller machine (ATM) offers a highly standardized service, while a tailor who stitches a fitting suit deals with heterogeneous demand. Some goods, such as cut flowers and milk are perishable, too. Some services are personal acting upon individuals (dentists). Some others work on facilities (cleaners), equipment (car repair), or data (insurance claim processing) (Wemmerlöv, 1990). Some services are highly customized and use a lot of labor, while others are standardized and automated (Schmenner, 2004).

2.9.2   Services as Collaboration

While IHIP focused on the nature of the material and immaterial products, the Customer Contact –school of thought emphasized the production process. Service processes differ from manufacturing processes by the simple fact that the customer has to be involved. The involvement can take many forms, simultaneous personal interaction in the same room, or over a telephone or through text. The service provider may interact with the customer in person, with customer's property or data about the customer. What matters is the personal, individual nature of the situation.

The customer-centric view of services is emphasized by a school of thought called Service Science. It is founded on a dichotomy between the traditional Goods-dominant logic (GDL) and the new Service-dominant logic (SDL) (Sampson & Froehle, 2006; Vargo & Lusch, 2004). GDL is based on a worldview where production and consumption are separated. Customers do not enter production premises. Products are shipped to markets and distributed through various channels. Customers are assumed to recognize the value of a product when they see and touch it, and consume it as they see fit. Marketing tries to capture and articulate needs and wants of the customers and communicate them to the producers the best they can. To the contrary, SDL is based on the conception that providers and customers collaborate to create customer value in a phase of production called co-creation of value. Therefore, GDL and SDL are developed based on different level of customer engagement in the production process. In other words, GDL assumes the customer is a reactor, while SDL assumes she is an actor.

2.9.3   Services as Resource Integration

IHIP and SDL have been combined into the Resource Integration Model (Moeller, 2010). (RIM), which sees them not as contradictory but overlapping and mutually reinforcing. Customer's resources are needs and want that justify spending, jobs to be done (Christensen, Hall, Dillon, & Duncan, 2016). Producer's resources are the facilities and the bundles of technologies (production function) that can get the job done. Services have four aspects: the customer's resources, the producer's resources, the service contract where the parties agree – perhaps after some bargaining – about a job to be done, and the service production process.

The IHIP categories are valid but apply differently to different aspects of a service. Customer resources are Heterogeneous, as individuals have varying needs and requests. Producer's resources are Perishable; without customers producers' capacity is idle and can't be stored for later use. The service contract is Immaterial, as services can be sold only as contractual promises of some future action. Inseparability appears in production processes where both parties must participate.

> RIM is an excellent example of segmentation. The IHIP -definition struggled with a conceptualization of services that included everything. Therefore the definitions could not be sharp. RIM split services into four parts, customer's resources, producer's resources, the service contract, and service production. Suddenly the picture cleared.

2.9.4   Services as State Changes

The third perspective to services is to look at what is the value that a service accomplishes. The value of service is a state change (Hill, 1977). Both material and immaterial entities have states. A while ago you had pain in your stomach; now you don't. The document that was a draft is now formatted and proofread. The state change must be seen as something else than the tools and processes that are needed to accomplish it. Taxi drivers sell rides, not cars. The basic feature of services is that they are processes that change states from something less valued to something more valued. A state change can be negative, that is, something that could have happened is prevented from occurring. The state change may apply to material goods and circumstances, such as repair and location, or to mental states such as education and amusement. The value of a service is founded on the difference between alternative states of things that can be attributed due to the service.

2.10   Types of Services

The perspectives can be combined to one definition. Services accomplish state changes by integrating customers' and producers' resources. Services can be segmented into types in many ways, such as whether the state change is in a material (repair, cleaning), or an immaterial (knowledge, skill) entity. The crucial classificatory principle is the extent to which a customer participates in the design and execution of production. This, in turn, depends on the level of standardization and specialization of the technologies employed. Self-service utilizes protocols that are standardized to the extent that customers can operate them without the direct involvement of a provider. The opposite is the service factory where services are prepared or executed without customer involvement. Between these extremes is an area of collaborative services. Collaborations are guided by, rules, rights, and responsibilities. The producer and the customer engage in bargaining about what, how, when, and how much, i.e., how an individual service is to be designed.

> Automatic teller machines (ATM) and internet banking are self-services, where the service provider establishes protocols, guidelines, and controls so that a customer can perform a process without personal involvement. A subway train can be considered a service factory as it follows its schedule regardless of how many, if any, passengers take a ride. Some health services do not require extensive collaboration (unconscious patient at ED), while continuous care of a chronic condition requires extensive collaboration and self-service. In medical cases where there are several possible options with different risk-benefit profiles, doctor-patient collaboration in clinical decision making is crucial.

The customer is conceived as an individual or a group of persons. A person is a human being with a bodily existence, a genetic makeup, a personal history, social relations, capabilities, and subjective preferences. In personal services, the person is the flow unit, i.e., the entity that is processed and whose state (location, health, mood) is changed. However, there are services where the flow unit is a property owned or managed by a person, e.g., a car, a lawn, or a house. The flow unit as property is individual in the sense that it is owned by somebody and the job to be done is defined by that somebody, while it is the piece of property that is subject to the service production process and the state change. Further, the person can be represented by a set of data. It too is individual as it depicts the state and circumstances of a person, but it can be processed as a case without the direct involvement of the person. Thus services can be segmented into personal services, property services, and care services.

> If you travel by air with luggage, you are a flow unit with three parts. At check-in, your person and your property (suitcase) are separated. You as a person are taken through security and boarding the aircraft where you are processed, i.e., your location is changed. Upon arrival, you and your property hopefully unite. Should that not happen, you file a complaint, and the matter turns into a case of lost luggage that requires tracking your luggage with the help of your data. In Healthcare specialists on diagnostics, radiologists and pathologists, treat patients as cases represented by data. A cardiologist is mostly interested in a cardiovascular organ system that happens to be the property of some person. Nurses and general practitioners often deal with persons.

2.11 Processes

Productive activities are conducted in time and space, therefore production can be described as one or several interlinked processes.

A process, generally speaking, is any set of transformations in time and space. There is (or was) a peace process in the Middle East. Preparing breakfast is a process. A production process is a special case. Division of labor and specialization require processes. A whole task is split into parts (steps), each is assigned to a specialized producer with specialized resources (workstation). When each concentrates on a narrow range of tasks, the dynamics of production can be observed; the best practices can be discovered, standardized and turned into protocols. Advantages of the learning curve can be realized.

A process is an organizational device that employs production functions repeatedly. A process can be run over and over again in a similar fashion to produce similar or near-similar results. A process description must include the following elements (Hopp & Spearman, 2011):

- Flow unit is the entity that is processed, i.e., its state is transformed
- Transformations (processing) are done by workstations that perform one or a few steps.
- The process is designed in advance as the route a flow unit follows from one step to the next (master setup).
- Before each step, the flow unit and the workstation must be set up (adjusted) to requirements.
- The flow unit is handed over from one workstation to the next.
- If the flow unit can't be processed immediately after the handover, it goes into a work-in-process inventory (WIP).
- A process precisely in time. Each step has takt time, each sequence of steps has cycle time, and a whole process has throughput time.
- A fully processed flow unit is throughput, which is measured as quality-adjusted throughput volume.

Production functions in operation are engaged in processing. The sequence of processing is called the value chain (discrete processing steps) or the value stream (continuous processing). In manufacturing, processing is additive – assuming that the process is under control and there are no or very few disturbances or quality issues that require rework – each processing adds value to the flow unit. In services, the value chain steps combine as multiples. A step can add value (>1), accomplish nothing (1), reduce value (<1) or destroy all accumulated value (0).

The transformations (state changes) are accomplished when a workstation does processing by exploiting technologies combined into production functions. A process, however, can't be equal to the value chain. It must be defined by setups, supported by logistic flows of supplies, such as parts, materials and energy, and preparations. The value chain (sum of processing) is in an asymmetric relation to setups and processing. Without processing, no transformations happen, and no value is created, but processing can't be done without setups, preparations, and supplies. These are worthless if processing does not happen.

> In surgical procedures, the value chain starts with the first incision and ends with the last stitch. Before that, there must be a plan for the procedure, adjusted to each patient (setup). The patient must be prepared, anesthetized and positioned on the operating table. If these are done promptly, but the surgeon doesn't show up, all effort spent on setup and preparation is wasted.

The conception that a process consists of the value chain and everything else is a waste (Muda) to be reduced or eliminated is misleading. For process improvement, the distinction between processing, setup, preparations, and waste is crucial. Processing can be improved through investments in technologies and skills. Setup and preparation are managerial issues of organization, i.e. how the Service Machine is constructed and operated. Waste is activities that do not contribute at all.

Setup and processing can be combined in many ways, giving a basic classification (segmentation) of process types.

– In a standard process, the setup is done once, then processing is repeated following the same setup. The cost of setup can be shared by a large number of transformations. The repeated use of setup is the central mechanism of mass production efficiency.
– In a formatted process, setup is done following a set of variables with defined ranges
– In a routine process, the setup is open to situational contingencies and bargaining between parties within limits.
– In an explorative process, the setup can't be done in advance, neither can the route of the flow unit be determined in advance. Setup is done once or a few steps at a time, then adjusted as information and experience accumulate.

2.12   The Business Model and the Service Machine

Business model describes how the business is planning to integrate key functions to make the profit and achieve a sustainable competitive advantage in the market. It has gained attention as people realized that the characteristics of the business are equally important in value creation and delivery besides the features of the product and services. Business Models are systematic ways to describe the entity of businesses, organizations that sustain themselves by getting jobs done and charging for the value / generating revenue proportional to the accomplished value. Business models integrate the demand side, what specific customer segments want to get done, and how they are reached with the supply side, how transformations are made to happen through the deployment of resources, the performance of activities together with supply networks.

A standard way to describe a business model is the Business Model Canvas (BMC) (Osterwalder & others, 2004; Osterwalder & Pigneur, 2010). It is a structured arrangement of answers to the following questions:

– What do you offer? What jobs can you get done?
– Whom do you serve? Who are your customers?
– How do you reach them?
– How do you relate to them?
– Which are your key resources and how do you structure them?
– Which are your key activities and how do you manage them?
– Which are your suppliers and other partners?
– What does it cost, and which elements cost how much?
– What is the source of your income? How do you get paid?

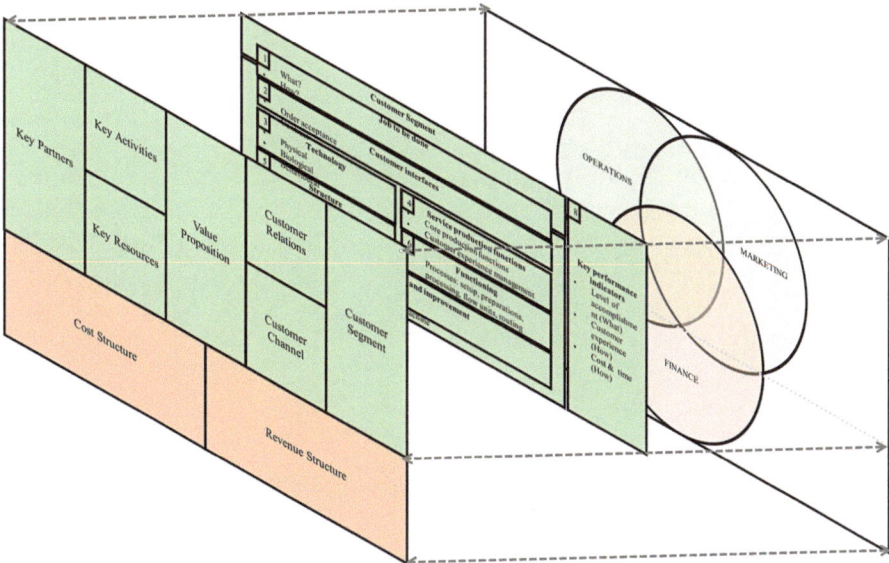

Fig. 2.2 Layers in the service business. (Adapted from the Business Model Canvas)

The key value of the BMC is to provide a structured and ontologically well-founded list of issues that any business must consider. The parts are related; if one is changed, it may affect the others.

BMC is the strategy part. The Service Machine is an elaboration of the operational parts of the BMC: the resources, the activities, the partners, and the producer-customer relations.

A Service Machine is a layer in a business system; it has to submit to the demands of the whole. It helps in understanding the way service production system is organized in a business. A customer interacts with the Service Machine to get the service produced. A Service Machine can evolve through trial and error, or it may be systematically and purposefully designed. Anyhow, a Service Machine must accept given value propositions, customer segments, technologies, partner ecosystems, and cost limits as generic design constraints.

The Service Machine could be seen as an intermediate layer as shown in Fig. 2.2. The top layer, BMC, provide the highest level of abstraction. In the bottom layer, the operations, marketing, and finance knowledge would work. The intermediate layer, Service Machine, provides the abstraction to understand the crucial components for a service production system. As the service production is co-creation, it has to integrate different concepts of Operations and Marketing with taking some cues from Finance.

Chapter 3
The Service Machine

Abstract We elaborate the metaphorical concept of Service Machine. An analogy between a Service Machine and a physical machine has been discussed with the help of a typical physical machine. A Service Machine is based on a fundamental demand-supply –constellation: what needs to be done and what can be done with the available means and circumstances. It must be deployed by users to get the job done. It can plan and execute service processes repeatedly upon demand. A template for designing a Service Machine is discussed along with some examples in service industry. The Service Machine template is a structured set of questions used to describe and analyze a service production system. This set of questions will guide a system designer to elaborate the description of individual building blocks of the service machine in a given context. The relationship between components of the Service Machine template is dynamic, i.e., the change in one of the component will affect other components. The Service Machines can be interconnected to form larger service production systems.

Keywords Service production · Customer segment · Customer interface · Production function · Demand-supply management

3.1 Universal Machine – Definition of Machine

We use the term Service Machine for a reason. Machines are systems, but not all systems are machines. A system can be the result of the evolution of living things, such as the breeding system of an animal (eggs, roe, seeds), a plant (a Banyan tree), or movements in the lithosphere that produce rivers and alluvial deltas. Social systems, such as marriage, kinship, and markets have evolved over generations without explicit designs. Crowds and queues form without anybody being in charge. Stock markets function by the collective participation of investors and regulators. In contrast to such systems, a machine is a system that is designed for a purpose that is defined by its output or functioning.

A machine has two or more functional components. A part belongs to an entity; a component performs a function. By this account, a knife may be built of several

© Springer Nature Singapore Pte Ltd. 2019
R. B. Roy et al., *Designing Service Machines*, Translational Systems Sciences 15,
https://doi.org/10.1007/978-981-13-0917-5_3

parts, but it is not a machine, while a pair of scissors is. The presence of components makes a system where the components interact.

When we say 'machine,' you think the steam engine and robot. We suggest, think of a bicycle and a rider. Technically speaking a bicycle is a compound machine made out of several simple machines, such as the wheel-and-axle, the lever, the screw, and the pulley (Sandori, 1982). Each simple machine exploits a physical phenomenon to achieve the mechanical advantage. A bicycle is a machine designed to put together several such mechanisms in a frame.

A machine transforms one energy form into another form for some purpose. For example, an electric motor converts electric energy into mechanical energy. The basic components of a machine are the energy source, transformation mechanism (engine), control unit and a working element that uses the transformed energy for a particular purpose. Engines are a subset of machines.

In a bicycle, the rider is the energy source, the engine, and the control unit. Muscle energy is converted through a pedaling motion through a chain and a pulley mechanism to the mechanical energy of the back wheel. The basic value proposition of a bicycle is that it amplifies speed to give the user enhanced reach. However, the user must submit to rules determined by the machine. The rider must learn to keep balance and use pedaling motions. Neither skill is obvious to a pedestrian and must be learned. A service infrastructure is also needed; there must be enough hard and plain surfaces to ride.

A Service Machine is a system that can plan and execute service processes repeatedly upon demand. The bicycle metaphor illustrates that the device alone can't create value. The machine must be deployed and put in motion by a user who has taken the trouble learning how to ride. Thus a Service Machine is based on a fundamental demand-supply –constellation: what needs to be done and what can be done with the available means and circumstances.

The basic architecture of a bicycle consists of the wheel(s), the powertrain, the frame, and paddle, as well as control mechanisms such as brakes and steering. A contraption that does not include these elements is not a bicycle. Within this architecture, innumerable designs can be conceived with various trade-offs. Each is based on a specific demand-supply constellation; what is the job to be done and what does it take to do it. Some riders want to go fast over smooth surfaces; a racing bike with a light frame and narrow wheels is designed. Some others want to lug heavy loads; a third wheel is added, and the frame is expanded to give carrying capacity. Still, some want to show off their motoric dexterity; a one-wheel circus bike is designed.

A bicycle has a physical existence, and a structural integrity even when not used. The bicycle-and-rider Service Machine is assembled for each ride; the engine is attached to the structure for each service instance. Without a rider a bicycle is useless; without a bike, there are no bikers. A service system, such as Emergency Medical Service (EMS), has its resources, ambulances, crews, and supplies scattered in various locations. They are assembled for each service instance.

Machines are usually thought of as mechanical pieces of equipment with parts that move. A machine, however, has also a broader definition. According to Merriam Webster Dictionary a machine can also be:

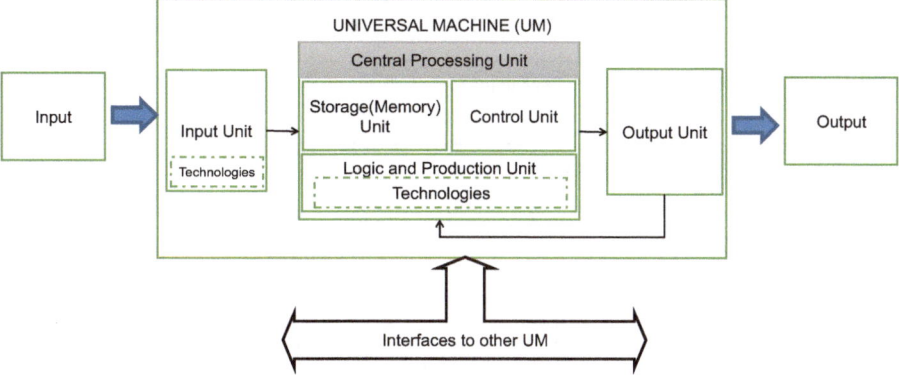

Fig. 3.1 The universal machine

- a person or group that does something efficiently, quickly, or repeatedly like a machine
- a combination of persons acting together for a common end along with the agencies they use
- a highly organized political group under the leadership of a boss or small clique

The concept "machine" can thus be used to describe socio-techno-economical systems that are designed and operated for a purpose.

The use of the term here follows a further definition found in Merriam Webster:

- a literary device or contrivance introduced for dramatic effect.

A prime example of dramatic use is the book "The Machine that Changed the World" (Womack, Jones, & Roos, 1990). It is about the automobile industry and the story of how the Japanese developed Lean Production. The "machine" here refers not to the automobile, but to the interconnected system of factories and supply chains that make efficient, high variety mass production possible.

The universal machine is depicted in Fig. 3.1. Basically, it is a black box that sits between inputs and outputs. Within the box, there are input units that receive, evaluate and approve inputs. The central control unit includes storage (memory) and control units that guide the logic and production unit, i.e., the technologies employed and bundled into production functions. The output unit consolidates and delivers the output to the interface. These are the universal aspects of all service systems.

3.2   The Service Machine

Services are supposed to benefit customers by changing the state of affairs to the better. The benefit can be the change in location (taxi), restored functionality (healthcare), new insights (education), and experiences (movies). The state changes are accomplished through technologies that the service producer organizes into

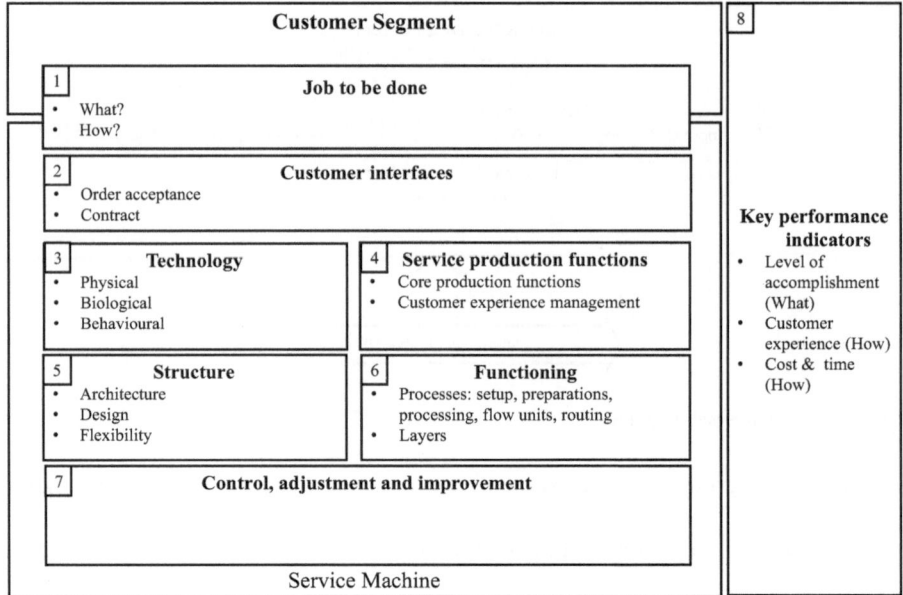

Fig. 3.2 The service machine template

production functions, bundles of technologies that make use of infrastructure. A taxi combines the technologies vehicle, driving, navigation, and fare metering, and operates on an infrastructure of roads and maps. The process defines tasks for the customer, such as telling the address and helping in navigation as required, as well as for the driver. In some services, the customer participates in person (taxi, haircut), through his/her property (car repair), or individual information (insurance claim). While some service processes can be unique and one-of-a-kind, most commercial services are repeat business. The Service Machine replicates the structure of the universal machine with the inputs being customer's initial state and the output customer's final state. The Service Machine is the structure of the resources, facilities, contracts, rules, information, and processes that make it possible to repeat a service process in similar or identical ways.

3.3   The Service Machine Template

The generic Service Machine is elaborated as the Service Machine Template in Fig. 3.2. The Service Machine template is a structured set of questions used to describe and analyze a service production system. This set of questions will guide a system designer to elaborate the description of individual building blocks of the service machine in a given context. The answers to these questions may lead to follow up questions to probe detailed view of the architecture of the service machine. This set of questions is being discussed below:

1. The job to be done;

   (a) What are you supposed to accomplish?
   (b) How are you going to do it?

2. Customer Interfaces

   (a) On which grounds and by what criteria are customer requests accepted (triage)?
   (b) What kinds of formal contracts or informal agreements are reached in order to co-create a service?

3. Which technologies, physical, biological, and behavioral, are employed?

   (a) What types of technologies are required to perform the job?
   (b) What are the technological options available to do the job?
   (c) Are the selected technologies appropriate to accomplish the proposed job?
   (d) Are the selected technologies make the production system economically and environmentally viable?

4. How are the technologies bundled into production functions

   (a) How are the technologies bundled for the core 'what' accomplishments?
   (b) How are the technologies bundled for the 'how' issues as customer experience management?

5. How is the Service Machine structured?

   (a) What are its overall architectures?
   (b) Is there any segment -specific designs?
   (c) Does it offer situational flexibility?

6. How is the Service Machine functioning in terms of processes?

   (a) How is the arrangement for setup?
   (b) How is the arrangement of preparations for individual cases?
   (c) How is the processing done?

7. How is the Service Machine controlled, adjusted, and improved?

   (a) What is the benchmark?
   (b) What are the bottleneck processes?
   (c) What are the control points?
   (d) What are possible options for improvement?

8. How is the performance of the Service Machine measured and evaluated?

   (i) How is the performance measured in terms of 'what' it accomplished?
   (ii) How is the performance measured in terms of 'how' it was done?

As with the Business Model Canvas, the assumption is that all elements are connected. If one is changed, others may change as well. A central question will be, if new technologies are introduced, how will structure and functioning be affected?

3.3.1   The Job to Be Done

What are you supposed to accomplish?
How are you going to do it?

Customers have needs and want that eventually turn into demand. Needs are universal in the sense that they relate to the human drive for survival, safety, belonging, and happiness. Needs can be assessed by a third-party observer using some generally accepted yardstick. Wants are expressed intentions to have or be something. Needs and wants often overlap, but not necessarily. A person may need but not want something (more exercise), or want but not need something (antibiotics for a trivial infection). Needs and wants turn into demand when backed up by purchasing power or membership rights (insurance, universal access to care).

When supply meets demand they – following the Resource Integration Model (RIM) – get integrated into a job to be done. The supply side has constraints from which follows that all demand can't be met as expected. In Healthcare the general supply constraints are technologies and production functions (all diseases can't be cured), patient's capabilities and motivation (healthy lifestyle requires effort), economic scarcity (resources may not be available at all, or not present at a specific time in a specific location), and the architecture and design of particular Service Machines (a patient may not be treated as connected flow but has to wait as patient-in-process (PIP) inventory).

Demand and supply engage in bargaining about how an individual service is designed, set up, and executed. From this follows that the jobs to be done may be different per segment and per individual. In services, the bargaining is about two basic issues. The What is about the expected state change, the How is about the process by which the expected state is accomplished. The "job" comes with a set of expected gains and pains. "What is to be done" explains the required state change. Each customer segment expects the state change in a specific way. The specific way could be articulated in terms of the expected gains and pains.

The requirement of the users of the online social network services is to get connected to the people of their interest. The "What is to be done" is to connect. By getting connected or through the process of getting connected, the people gain lots of social, professional, and personal values. It helps them to reach out their relations in far off places and interact with them. To get connected using the online social network, the people need to spend their valuable time and money, share information, and more importantly identify the relevant information by managing the information overloading. The customer of an online social network service would like to get connected (what) to make and foster relations easily without sacrificing their resources and privacy.

There can be trade-offs between "What", and the "How," some effective thera-pies may necessitate pain and inconvenience. The customer part in a Value Proposition Canvas is a good tool to represent the job to be done. The service machine is expected to raise the gain and reduce the pains of the customer.

3.3.2   Customer Interfaces

On which grounds and by what criteria are customer requests accepted (triage)?
What kinds of formal contracts or informal agreements are reached in order to co-create a service?

Services are transformations to which customers and producers participate. The rules of engagement are determined in the service contract, the user manual of a service machine, which stipulates the tasks, roles, rules, rights, and responsibilities of the parties.

All service requests can't be accepted by a Service Machine. A gatekeeping function sorts arrivals into acceptable and rejected. In Healthcare triage is used to segment demand in terms of urgency. One of the main item to be included in the contract or the agreement is to incorporate the local social, legal and ethical consid-erations. It is an implicit assumption that all legal services would not harm the cus-tomers intentionally. The customers may or may not aware of all the risks involved in the service. It is the service provider's obligation to inform the customer regard-ing the risks. It is also important for the service provider to inform the customer appropriately if something went wrong. The interface for informing the customers depends on the ethical, cultural, social, and technological state of customers. If the interfaces for the service machines are not designed properly, it may result in high customer dissatisfaction. When bargaining is concluded, and a service agreement has been reached, a service contract is established, explicitly or implicitly. In addi-tion to that, it is customer's responsibility to use the service machine responsibly and without harming the system. Like any physical machine, the user characteristics for the service machine can be defined and enforced. Similar to insisting on the minimum age for using a motorbike, some of the service machines insist that the user should be an adult to use the machine responsibly.

The service types mentioned above can now be elaborated using the setup – pro-cessing distinction.

The service factory has a fixed structure and setup, that is communicated to cus-tomers in advance as the value proposition. The service, such as a metro train, fol-lows a given schedule. The customer's options are – in normal cases – to take it or leave it. Automatically renewed subscriptions are service factories where actions are initiated by a pre-planned schedule.

Collaborative services require bargaining, i.e., finding out a workable combina-tion of what needs to be done and what can be done. Service systems typically reduce the heterogeneity of demand by offering service menus and limiting the bargaining space to pre-defined options, i.e., formatted setups are used. Collaboration becomes more intensive in situations where pre-defined service menus can't be used. This is typical in specialist care, such as complex cancer, where the process can't be planned in advance from end to end. Bargaining and setups are required at each step.

Self-service make use of highly structures interfaces and service user manuals. A provider offers a template, resources, and instructions, the customer initiates and performs the service as pleased.

The customer interfaces need to be designed based on the type of the service production- service factory (almost closed), collaborative service (partially open), and self-service (maximally open).

> Most of the online social networks behave like a self-service. The customer operates the service machine by accepting the terms and conditions (manual). Thus the users accept to provide some information to the provider and allow the machine to communicate with the user ensure the services are running properly. Many times, the machine provides the user, complete control over the information to be provided and used. If it is not clear about what and how information needs to shared and used, it may disrupt the functioning of the machine (legal battles regarding breach of local law) as well as may harm the user (compromise on privacy and more). By accepting the agreement, the customer agrees to use the machine in a defined fashion. Many providers require the users to have minimum age criteria to use the service. The providers through their contract accept not to harm the customer and respect her privacy and protect her dignity.

3.3.3   Technology

> Which technologies, physical, biological, and behavioral, are employed?
>
> (a) What types of technologies are required to perform the job?
> (b) What are the technological options available to do the job?
> (c) Are the selected technologies appropriate to accomplish the proposed job?
> (d) Are the selected technologies make the production system economically and environmentally viable?

Available technologies are resources that can be held by both customers and providers. Technologies build on knowledge about how the world works, and can be physical, chemical, biological, information, and behavioral. Many times in services, the customer resource is information. The technologies used by the providers should complement or supplement the resources of the customer in order to facilitate the resource integration. As service production is a co-creation activity, it is important to engage the customer in the production. The technology is selected based on the characteristics of the customers and their requirements.

> In online social network services, the connections only happen if the people who are related to the users are also actively using the services and they share the information to identify the relations. Thus the service providers commonly use the behavioral technologies like hooking to engage the users. In addition to that, it uses the information technology for the recommendation and sharing appropriate information. With the advent of technologies such as mobile technology, imaging, and video technologies, there are alternative ways of getting connected. Primary technologies involved are information and behavioral. The naive

technology for managing the large information is to make use of human capital. But, with the large volume of information using human for handling information is impractical and insecure. Thus, the technologies like artificial intelligence are used. More importantly, it is necessary to verify the economical and environmental feasibility. Developing a social network platform which could be accessed only through a high-cost gadget may not be a good idea if we are designing it for the mass market. Similarly, if we are developing a platform where a large amount of data needs to be transferred, it results in huge bandwidth cost which may not be affordable to all. Probably, as it is an online platform, it may not affect the environment directly. But it is necessary to understand how big the information would be and how huge will be its impact on global warming due to the data servers it should possess.

3.3.4   Production Functions

How are the technologies bundled into production functions

(a) How are the technologies bundled for the core 'what' accomplishments?
(b) How are the technologies bundled for the 'how' issues as customer experience management?

Production functions are bundles of technologies, including customer actions that accomplish state changes.

Consider the collection and analysis of a blood sample for diagnostic purposes. Blood analysis is founded on biological knowledge about the properties of blood (red and white cells, plasma, etc.) and what it may tell about a patient's condition. A sample is collected using a simple machine syringe, a reciprocating pump consisting of a piston that fits within a cylindrical barrel. The piston exploits vacuum to take in and expel liquid through a discharge orifice at the open end, which is fitted with a hypodermic needle. The production function is organized as a process that includes the steps a doctor ordering blood analysis, instructing the patient to go to a collection point, identifying the patient, collecting the blood in a vial and labelling it with patient ID, sending the sample to an analyzer, recording the results and delivering them to the doctor. The accomplished state change is new knowledge about the patient's status.

Service production functions have two aspects. The core is the What, i.e., the technologies that enable a state change. Since the customer is involved and brings resources, customer experience management, the How's, is an essential part of the production function.

The social network connects (what) the people to their relations. Thus, it allows the users to change their state of "being in the silo or missing someone" to "socially connected." It is important for the provider to make the connections easy, secure and safe. The user should feel that they are safeguarded from exploitation and could communicate with relations easily. Overloading the users with irrelevant data may take away the users. Thus the service production function should be designed to avail the relevant information.

3.3.5 *Structure*

How is the Service Machine structured?

(a) What are its overall architectures?
(b) Is there any segment -specific designs?
(c) Does it offer situational flexibility?

The production functions accomplish state changes and accumulate value. They, however, can't exist in a vacuum, but need to be organized and structured by organizational architectures and designs, and fine-tuned in bargaining about jobs to be done.

Architecture refers to the basic elements of production and how they are connected as structures. Buildings are architectures set in stone, concrete, and steel. In politics, a constitution is an architecture protected by the cumbersome procedure. Product architectures follow the intended performance, production architectures follow from the production function.

> A passenger car has an architecture based on four wheels and a compact chassis, which differentiates it from two-, and three-wheeled motor vehicles. A dairy has an architecture that differs from a garment factory. A restaurant architecture includes tables, kitchen, and an order-delivery system. The architecture of travel includes a departure location, a destination, a route, a transport device, and the cargo to be transported (the flow unit with a job to be done).

In short, organizational architecture refers to purposes, assets and their relations. Service architectures include customers.

> The architecture of the social network is that the service should connect two people. Irrespective of the medium, the users need to communicate, share information (photo, video, news, and gifts), involved in activities together. The architecture includes people and a medium.

Design means how components and tasks are organized within an architecture. The Business Model Canvas (BMC) depicts the basic architecture of a business. Within these constraints, organizations can develop individual designs. Designs have stability, but rearrangements can be made with some effort. Passenger cars can be sized and styled for different customer segments. Architectural change implies investments and/or major upheavals. Designs can be changed with less effort.

> Restaurants have three basic designs, buffets from where customers pick different foods; a counter where orders are taken and delivered, and seating and service at the table. Discrete manufacturing has three basic designs, the job shop, the disconnected line with buffer work-in-process inventories, and the connected flow (moving assembly line). In case of an online social network, the design could be broadcasting (like Twitter, Mastodon), one-to-one (like Messenger, Skype), and many-to-many (like Facebook) communication.

A service system can be designed to be a single-purpose Service Machine that accepts only one type of input and performs a fixed set of operation. All arrivals that do not fit the description are rejected, and no bargaining is allowed. Service systems can also be designed for flexibility; a broad range of different cases can be handled. In such cases, the Service Machine needs to have elaborated methods for bargaining about how individual service instances are defined and formed. A system that with the same design and constant assets can answer to a range of requests is flexible; flexibility, in turn, means that the Service Machine can change into, and alternate between different operating modes.

A rural health center in an isolated area is a Service Machine with fixed and limited assets. Being the only local unit for medical help, it has to deal with all types of demand, from open wounds to pregnancies to abdominal pain to advise on nutrition and hygiene. From patient to patient, the Service Machine has to switch operating mode from emergencies to diagnostic exploration to prevention without the possibility to call in new assets. In an online social network, the flexibility is at multiple levels. The service could be a centralized (Twitter) or a distributed (Mastodon) one. In addition to that, like Facebook, it provides the users to select the type of communication. The flexibility should be there to accommodate the requirements of a community also. For example, some of the content shared in social media is appropriate for a particular age group, country and so on. By selecting the options, the customer should be able to customize the same.

3.3.6   Functioning and Processes

How is the Service Machine functioning in terms of processes?

(a) How is the arrangement for setup?
(b) How is the arrangement of preparations for individual cases?
(c) How is the processing done?

A Service Machine at work can be observed as one or several processes going on. Processes are the engines and exhibit the dynamics of the Service Machine.

In an online social network, the user needs to get registered to make use of the service machine. Only the users who are eligible to use the service machine are allowed to get registered based on the information provided. The machine enables the users to get connected and communicate each other. It also enables the users to break the connection, if necessary. Furthermore, it allows the users to be actively (always logged in into the network) and passively (occasionally logged in) connected to their relations, and even abandon the service machine.

In services, the customer is the flow unit to be processed. The customer can appear as a person, property, and information. From this follows that a Service Machine can simultaneously run several processes, which can be imagined as process layers.

In emergency and specialist inpatient care the patient process proceeds at several layers. There is a (a) diagnostic process that aims at creating a correct understanding of the medical problem and its causes. The diagnostic process is fed by data from (a) data collection process of discrete information acquisition through blood and tissue samples and images, and (c) a continuous monitoring process. These build the foundation of the (d) clinical decision-making process where accumulated data is judged against medical knowledge. The clinical decisions feed the (d) procedure process, i.e., the clinical interventions directed at the source of the problem, such as an injury or a failing organ. The patient as a whole person is part of a (c) care process of basic nursing, comfort, food, connections with relatives, food, etc. Here the layers are identified according to the flow unit: the patient as a case represented by information, the patient as owner of 'properties' such as organs, body parts, and their problems, at the patient as a whole person with basic needs, a personal history, social relations, values and preferences.

In an online social network, mostly the information of the user is processed. Although the service provides a personal experience, the property of the user, information is processed and shared.

In patient processes, the patient is the unit of analysis, and the process consists of a journey through several steps. Processes can also be described with a workstation or step as the unit of analysis. A workstation, such as a facility for plastering broken bones, receives a stream of patients, makes a setup to select and adjust the procedure, performs the step (processing), hands the patient over to the next step, and then receives the next patient, over and over again.

3.3.7 Control, Adjustment, and Improvement

How is the Service Machine controlled, adjusted, and improved?

(a) What is the benchmark?
(b) What are the bottleneck processes?
(c) What are the control points?
(d) What are possible options for improvement?

A process is a coordination device that arranges processing in space and time. Productive resources need to be scheduled, i.e., prepared and assigned time-slots. The scheduling can't always be done with sufficient precision, because of all factors of production exhibit variability, nonuniformity of a class of entities or random deviations from the expected. Variability is the lack of pattern or predictability in events. When production systems have been designed, daily management is about dealing with variability, to find it, avoid it, amend it, and assure against it.

Any deviation from target or missed deadline creates a cascade of variability within the system. Internal variability shows up as waiting time, excess inventory, or excess capacity. Other types of variability are unscheduled downtime, tool breaks, absenteeism, and malfunctioning.

Knowledge of production dynamics helps to drive out variability through standardization and control. Where this can be done, production systems operate like clockwork. In personal services such as Healthcare, this can't be done. In an imperfect world, variability must be accepted and managed within the limits of the possible.

Process improvement usually requires drawing a map about a process, examining it in detail, and identifying issues to be changed. Mapping creates knowledge of a process and follows the general pattern of knowledge development.

A process has an ontology (what is it) that can be described as inputs, outputs, steps, and connections. This can be done by examining an original design (if it exists) and conceptions of how the process should run, by asking involved parties how they think it works, or by third-party observers. This creates a conceptual model of a process depicting its elements, such as steps, flows, and performers (workstations). A process can be described empirically (epistemology) by observing, counting and measuring what happens. This creates a quantitative model that illuminates the arrival rate of flow units; the capacity utilization rate of workstations, inventory and throughput volumes; and resource and time consumption. With data, process dynamics can be explored. Typical dynamics are bottlenecks, i.e., workstations that can't keep pace with others, thereby accumulating inventory and starving the next steps from flow units.

> The benchmark for the online social network would be the real world network. The services need to easy and intuitive for a social human being. The adjustments need to be done to prevent unnatural and unsafe experience. In every level, the improvements are required engage the users to operate on a virtual platform rather than in a real intuitive platform. In real life, the users can understand and communicate with others easily. But in social media, the knowledge of technology is expected to maximize the user experience.

The control and management of the service machine for the social network could be centralized and proprietary (Facebook, Twitter, Youtube) or decentralized and open source (Mastodon, Peertube). In centralized proprietary services, the improvements are made based on the perceived customer satisfaction and investors interest. In open source social network, the improvements are mostly community driven. In case of centralized proprietary services, the research and development were done to improve the customer satisfaction, but the customer may not be able to modify the working of the system. But in open source machines, the customers can force the machine to modify.

3.3.8   Key Performance Indicators (KPI)

> How is the performance of the Service Machine measured and evaluated?
>
> (a) How is the performance measured in terms of 'what' it accomplished?
> (b) How is the performance measured in terms of 'how' it was done?

The performance of a Service Machine is measured and evaluated in terms of how well it gets jobs done. The leading key performance indicator is the level of accomplishment. In services, it is the outcome of a string of processing that adds or reduces the expected state change. As customers participate in service production as flow units, their experience needs to be registered. Particularly in personal services with interactions, employee experience should also be registered.

> The connectedness is one of the key performance indicators for a social network. The connectedness depends on the number of users of the social network services. The measure of

connectedness could be degrees of separation. As the number of users get connected in a social network increases, the degree of separation decreases.

Value, strictly speaking, is the relation between accomplishment and expense, what you get in relation to what you give in terms of money, time, and trouble. Therefore the KPI must include measurements resources expended (cost), and time spent. In contrast to accomplishments, there are additive: cost and time consumption can't be eliminated, they can only accumulate.

3.4   Service Systems as Connected Service Machines

The Service Machine is a systemic construct. That means Service Machines can be conceptualized at various system levels (units of analysis), say a General Hospital, an Outpatient Clinic (OPC), or a procedure, such as taking and analyzing a blood sample. Service Machines as components of larger systems require interfaces with other machines, which can be envisioned as exchanges of inputs and outputs. An OPC sends a service request to a blood analysis Service Machine, which then delivers its output to the OPC.

Service Machines typically operate within infrastructures. An ambulance service depends on roads and communication devices. The Service Machine – infrastructure interfaces are typically codified, and function akin to self-service – the infrastructure is available, the Service Machine uses it at will.

Chapter 4
Case Studies

Abstract We applied the Service Machine template to the analysis of a rather complex service system, Emergency Medical Care. We analyze two Service Machines in Emergency Medical Care, namely Emergency Medical Services (EMS) and Emergency Department (ED). EMS and ED are primarily focus on saving and stabilizing the life of a patient who requires immediate medical care or advice. EMS is designed to provide pre-hospital care and transport the patients who require immediate action to save and stabilize their lives to a hospital for definitive care. The role of ED starts immediately after the EMS and the core service functions are: stabilize, diagnose, care, and guidance. These two Service Machines are interconnected and there is some overlap in their service production functions. The components of Service Machine Template for these machines are discussed in detail. A service blueprint of EMS and ED has been presented. Later, we discuss the similarities, dissimilarities and inter-connectivity of these machines.

Keywords Self-service system · Collaborative services · Service factory · Emergency care · Service blueprint · Save and stabilize

Emergency Medical Services (EMS) and Emergency Department (ED) are service machines primarily focusing on saving and stabilizing the life of a patient who requires immediate medical care or advice. These two service machines have overlap over the objectives, but they are different in operations as depicted in Fig. 4.1. The customer segments for the service machines are different, but the same segment may use both the service machines. Emergency Medical Services (EMS) is designed to provide pre-hospital care and transport to the hospital for the patients who don't have access to the definitive care and require immediate action to save and stabilize their lives. The role of Emergency Department (ED) starts where the part of EMS ends. The ED provides definitive and advanced care including surgery which is not usually possible in ambulances. ED provides services to another customer segment who directly walk into the service point. That customer segment identifies the need for urgent medical care or advice but may have easy access to the hospitals.

The customer segment of EMS doesn't have easy and quick access to ED. EMS primarily acts as access provider to medical care that is available at ED.

© Springer Nature Singapore Pte Ltd. 2019
R. B. Roy et al., *Designing Service Machines*, Translational Systems Sciences 15,
https://doi.org/10.1007/978-981-13-0917-5_4

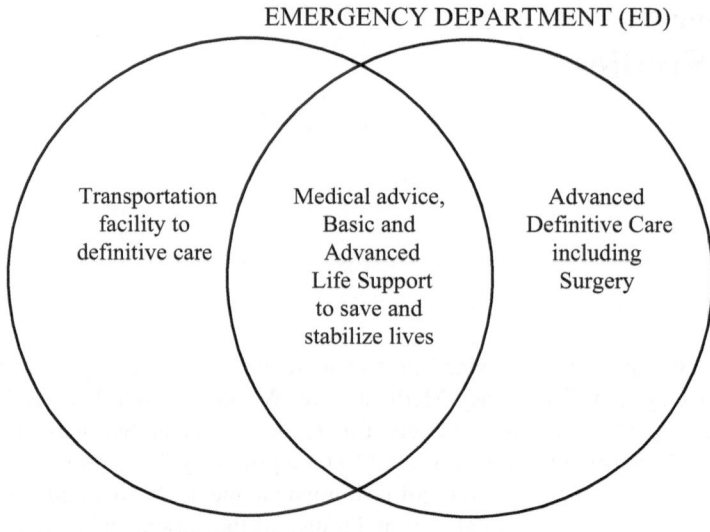

EMERGENCY DEPARTMENT (ED)

Fig. 4.1 Services provided by Emergency Medical Services and Emergency Department

4.1   Emergency Medical Services (EMS)

An incident that leads to an emergency situation can be structured using the "Star of Life" (National Highway Traffic Safety Administration, 1995) model with the six steps:

1. Early Detection
2. Early Reporting
3. Early Response
4. On-scene care
5. Care at transit
6. Transfer to a definitive care

Thus the user of the EMS detects the need for immediate medical intervention and report the same to the service provider who offers pre-hospital care and transport to an emergency department (ED) as a service. Once the incident reported, the provider has to evaluate the situation and respond to the request appropriately within a given time-frame and an available resource, i.e., an ambulance. Once the ambulance reaches the location, the ambulance crew provides on-scene care, if possible. If on-scene care can't solve the issue, transfer to a location with the appropriate resource, i.e., a hospital Emergency Department (ED); and when required, care given to a patient while in transit.

As a service machine, EMS operates on a patient who requires immediate care to save and stabilize her life. It transforms a patient in a medically critical state to a saved and stabilized state by means of on-scene, or pre-hospital care, i.e., a

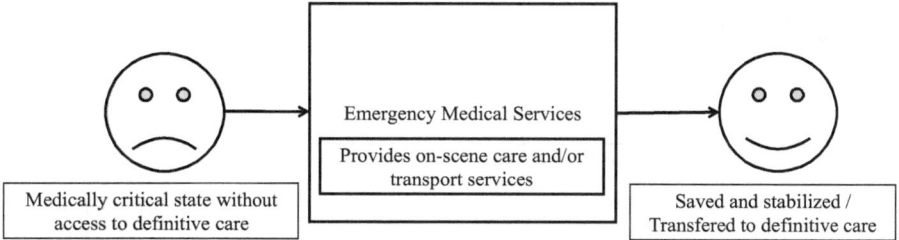

Fig. 4.2 Service machine for Emergency Medical Services

production function that comprises technologies, such as resuscitation, bandages, tourniquets, and painkillers. The second or alternative state change is the change of location as shown in Fig. 4.2. The EMS Service Machine is designed to provide pre-hospital care and transport services to customers.

EMS constitutes of two sub-services – the dispatching services executed through the call center and the ambulance services. To the ambulance crew, the customer is a patient in person (with some damaged bodily and mental properties), to the dispatch center the customer is a case represented by data.

4.2   The Service Machine Template for Emergency Medical Services

In EMS the customer is best described as a consumer or a user, since EMS in many countries is a government or charity -funded service that does not include bargaining between buyer and seller. A patient is, per definition, a person who has requested the services of Healthcare. Thus the EMS 'customer' is a patient without direct access to definitive care which requires immediate care and/or transport to definitive care to stabilize her condition and save her life. The service machine executes the "job to be done" for the patient. The service template for EMS is as shown in Fig. 4.3.

4.2.1   Job to Be Done

The Service Machine implements the value proposition offered by the provider to complete the job to be done for the customer. Thus a service machine matches the offering (supply) with customer's requirement (demand).

The emergency patient comes up with the job to be done (what and how), avoidance of immediate death (mortality) or permanent injury (morbidity). There are around 32 chief medical complaints which could result in mortality and morbidity. Immediate medical intervention is required to save the life or to avoid morbidity.

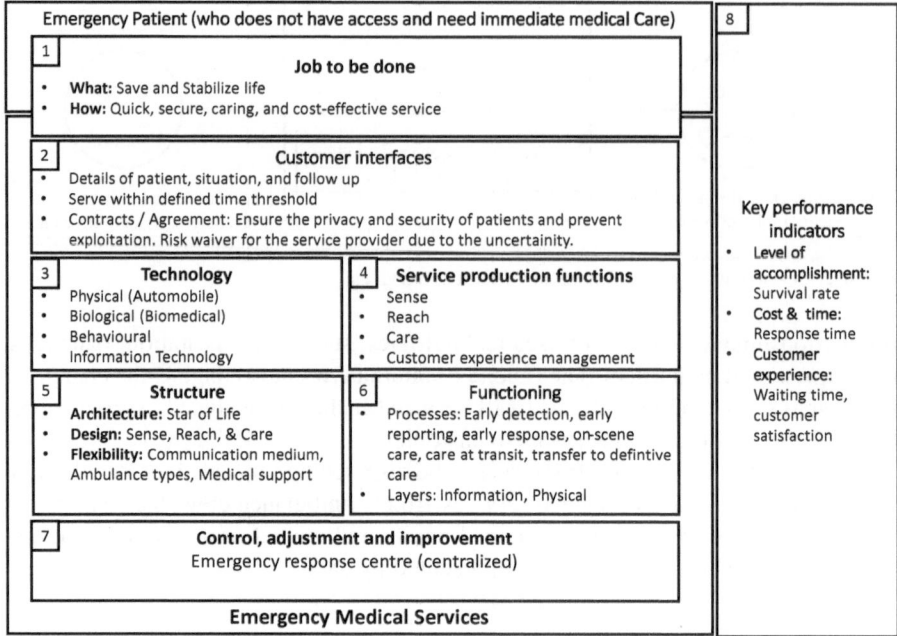

Fig. 4.3 Service machine template for Emergency Medical Services

The patient comes up with the requirement of gains and acknowledges that there are some inevitable pains from and during the services. The patient needs to go through events that involve uncertainty (waiting for the ambulance to arrive) and mental and physiological stress. The recent advancements in technology could help in personalizing the services, and the customer wishes to have personalized care. The services need to be provided to the patient securely which have to take care of the privacy, safety, and comfort of the patient.

As a payer of the service, a customer looks for the cost-effective solution. As most of the emergencies are a possible avenue for exploitation, it is vital to have cost-effective and quality services.

4.2.2 Customer Interfaces

All production systems exhibit variability. In services, customer idiosyncrasies constitute a significant source of variation that will impact performance. A Service Machine is designed for specific purposes. Therefore it needs to distinguish the customers it can serve from those that it can't or those who would be served better elsewhere.

> EMS is designed for rapid response. For anybody with a medical problem, quick response is preferable to waiting. The rapid response capability, however, comes with a cost. EMS is expensive, as they require 24/7 availability, advanced equipment, and slack resources on standby. Therefore access to EMS must be restricted and a gatekeeping function employed.

The patient's request over telephone would be scrutinized (triage) and evaluate the priority using information and medical technology. Based on the triage, the customer would be segmented. As per the scope of EMS offering, the services would be provided only to the patients whose health state is medically critical.

The patient is expected to provide the appropriate information to the provider to understand the requirement. Based on the information provided by the patient, the services would be either render or denied by the dispatcher in emergency response center. The reasons for denial of services could be due to demand (the patients does not require immediate care) or supply (unavailability of ambulance or ambulance crew, outside the service catchment area, and so on).

By initiating the services, the user of the services abides by the implicit agreement that they would be cooperating with the service provider to create value. The agreement alleviate the risks for service providers as there are lots of uncertainty in the outcome of services, especially in an emergency.

> The state of a patient would be so critical in many cases. It would be difficult for the service providers to save the life of the patient using the known knowledge. The providers could do whatever they are supposed to do, but saving the life may not be possible. It is essential to protect the services providers from ill-effects due to such risks so that providers could take up the requirements without hesitation. If providers are exposed to such ill-effects, there is a probability that the providers may refrain from attending complicated medical cases.

At the same time, the contract or agreement with the payer enforces a quality service to the customer with the defined time frame. The services need to be within the legal and ethical foundations of the country. The agreement helps the customer to prevent the exploitation.

4.2.3   Technology

The available technologies would be used to match the value proposition offered by the machine with the customer's job. Different technologies such as information, automobile, medical, and behavioral technologies are used in EMS. The information technology is vital for communication between the provider and customer, and internal communication of provider. The automobile and transportation technologies help in providing the ambulance services appropriately. The medical and behavioral technologies are required ensure appropriate medical services and patient's mental and physiological support.

Table 4.1 illustrates how the supply side has applied new technologies to meet customer requirements better. The urgency –imperative calls for speed in all steps. For instance, the selection of smaller toll-free number for emergency call helps the customer to remember the number and dial it easily.

Table 4.1 Technology available for the job to be done in Emergency Medical Services

Customer requirements/demand	Description on the value creation	Technology
Want to get in touch with an actor who can provide advice and help in an emergency, and communicate the urgency and the details of the situation and its location	Enhance access, improve communication, and support co-design of the service instance.	Introduction of nation-wide or state wide toll free number with 3 or 4 digits
		Rich communication content through smartphones, pictures, videos and GPS maps to determine exact location.
Follow the progress of the case.		Algorithms and decision trees for quick and accurate assessment of urgency.
Want to get quick access to medical care	Depends on the urgency and accessibility to the location of incident various technologies are evolved. The logic for service is decided based on the interaction between the customer and provider. These services are offered by the providers to enable faster and appropriate access.	Scoop and run:
		Various ambulance vehicles, such as vans, jeeps, helicopters, motorbikes, tricycles, boats, etc.
		Real-time information of ED capacity and preparedness. Advanced notice on incoming patients and their condition
		Stay and play:
		Ambulances with doctors and specialized equipment, such as trauma kits, defibrillators, etc
		Directing local doctors to the scene[a]
		Use drones to carry supplies[b]

[a]Silverston, P. P. (1985). Physicians at the roadside: Pre-hospital emergency care in the United Kingdom. The American Journal of Emergency Medicine, 3(6), 561–564. https://doi.org/10.1016/0735-6757(85)90172-X

[b]Webredactie Communication. (2014). TU Delft's ambulance drone drastically increases chances of survival of cardiac arrest patients. Delft University of Technology. Retrieved from https://www.tudelft.nl/2014/tu-delft/ambulance-drone-tu-delft-vergroot-overlevingskans-bij-hartstilstand-drastisch/

Understanding the customer and translating it into specifications probably the easiest way to identify the technology. The Quality Function Deployment or House of quality could be an excellent tool to determine the appropriate technology required for meeting the requirements.

The selected technology should be economically viable. For example, using helicopter ambulance may be one of the fastest ways of transportation. The cost involved for its operation and required infrastructure (helipad, security system) is too vast, and as a result, it may be able to serve a small number of people. Thus such technologies are economically challenging. In addition to that, the transportation facilities require more and more land for expanding roads and associated infrastructure. Unscientific use of land will adversely affect the environment. Thus for selection of technology, the economic and environmental aspects to be considered.

4.2.4   Service Production Function

Once the provider accepts to provide the service, the core service production function comes into action within the Service Machine. The available technologies would be bundled to form the service production function(s). Depends on the availability of the technologies the service functions would be improved or modified.

> In EMS, the core service is the pre-hospital care and transportation. The service engine makes use of technologies such as IT, biomedical, automobile, and behavioral technologies to operate on the different customer flow units. The patient could be flowing through the service machine as a person, data, and a property. The service machine needs to be tuned appropriately to operate on these service flow units.

The service production in EMS includes three primary state changes and these state changes are brought by the service production functions: Sense, Reach, and Care. In the Sense service production function, the customer informs the details regarding the situation at incident location and provider would sense the need. At this point, two cognitive transformations happen. The customer would transition from uninformed to informed state, knowing the availability of services. The provider similarly transfers from an uninformed to an informed state; a service request is received and evaluated to make decisions whether the ambulance services would be dispatched to the scene. Reach is the production function of moving and navigation. Care is the production function of giving care at the location and in transit. The resources, technologies, and the state change involved in service production function for EMS are depicted in Table 4.2.

Table 4.2 Resources and technologies in service production function for Emergency Medical Services

Production function	Machine resources involved	Technologies	State change in machine	Role of customer	State change in customer
Sense	Call center	Information, behavioral, bio-medical technologies	Acknowledge	Understanding the situation and report the same	Access to the service
Reach	Call center Ambulance	Information, behavioral, bio-medical, automobile technologies	Order acceptance	Provide necessary information for evaluation of the scene over phone and wait (time) for the arrival of ambulance	Availability to the service is received
Care	Call center Ambulance Medical Team	Information, behavioral, bio-medical, automobile technologies	Processing	Collaborate with the ambulance crew on-scene and undergo treatment	Saved and stabilized

4.2.5 Structure in EMS

In a typical EMS, the customer as a user and customer as a payer is different. Usually, the customer as payer is the government which outsources the work to a service provider or allocate the work to a department of government. The customer as a user could be the patients him/herself or a person who makes the call to EMS for services. The payer would not involve in the process level operations while the user plays a vital role in co-creating the value during each service. The primary inputs from the user are information and the user itself. Fig. 4.4 depicts the three agents in a typical EMS system where the system is funded by Government. In this case, the Government would act as the principal and service provider would act as the agent for Government to provide the services to its citizens. The government could monitor the working of EMS as in modern EMS systems have most of the activities are captured using Information Technology (IT) systems. The systems where EMS provider is paid by the patient directly would have a different structure. In those cases, the payer would be patient him/herself or insurance provider. In every case, the underlying relations between the agents remain more or less same.

The core service functions are pre-hospital care and transportation using ambulances (supplemented with Fire and Police as per requirement). The procedures followed during the services are governed by standard medical protocols. The user needs to play the role of provider of information and input to the medical procedure as him/herself. The involvement of user and service provider during six phases of Star of Life concept used in EMS varies. The effort required to be put by the user in

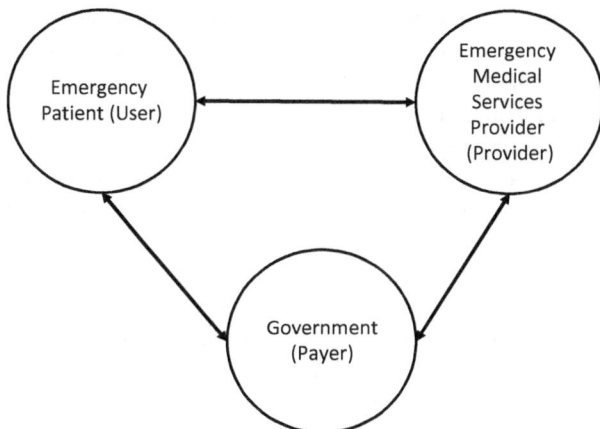

Fig. 4.4 Agents in EMS system

Table 4.3 Work allocation, interaction, and verifiability in Emergency Medical Services

Phase	Process	Work	Interaction	Verifiable consumer actions	Verifiable provider actions	Service production functions
Detection	Standard	Consumer	Low	No	NA	Sense
Reporting	Standard	Consumer	Low/ medium	No	Yes	
Response	Standard	Consumer – provider	High	No	Yes	Reach
On-scene care	Standard	Provider	Low/ medium	Yes	Yes	Care
Care in transit	Standard	Provider	Low/ medium	Yes	Yes	
Transfer to definitive care	Standard	Provider	Low/ medium	Yes	Yes	

the initial stage is high, and it decreases as the stages advance. At the same time, the effort required to be exerted by the service provider increase as the stages of services advance.

The work allocation, interaction, and verifiability, in the EMS service, are shown in Table 4.3. We have three core service production functions in EMS: *Sense, Reach, and Care*. In stage *Sense*, the user detect and report the same to the provider and provider could sense that there is a requirement. Then the user and provider interact and identify the actual requirements and allocate the resources accordingly in *Reach* stage. The major processes in reach phase are triage the patient and dispatching of the appropriate resources. In the *Care* stage, the provider provides the proper care to the user.

Depends upon the catchment area for the service, the service providers need to have appropriate organizational structure. The organizational structure is important as it plays a crucial role in control and management.

In India, National Ambulance Services Project runs by National Health Mission (NHM), aims to provide nation-wide EMS. Although it is a national project, the implementation is done in state level. Each state is responsible for implementing the EMS in their state and NHM would financially support the projects. The administrative structure should be such that it should allow managing the ambulance fleet easily and responsibly. The structure needs to respect the federal structure of the country. An ambulance requires a pilot (driver) and an Emergency Medical Technician (EMT) for normal working. The unavailability of any one of the crew member makes the ambulance off-road. Thus, each ambulance is allotted with three pilots and three EMTs who work in pairs on shifts and entitled to weekly off. The EMTs are a select cadre of paramedics with training on the van for 6 weeks or shorter length. The training is multidimensional and covers institutional, hospital and ambulance components. The EMTs on the van are supported by the doctors at call center through the telephone. Every 15 ambulances are managed by one operation executive and one fleet executive. The operation executive supervises the patient care, and the paramedics and the fleet executive look after the vehicle care and the driver. This pair of executives reports to a district level manager and an administrative officer. These district managers are managed by a zonal executive manager. The zonal managers report to State level executive manager who reports to a consultant from NHM, national level for the project. The States are free to adopt the structure that suits its working. The structure adopted in India is shown in Fig. 4.5.

Thus while defining the structure, it is important to understand the operational requirements.

4.2.6 Functioning and Processes

The overview of processes involved in transforming a medically critical patient to a saved and stabilized state is shown in Fig. 4.6. The initiation of the services depends upon the knowledge, skills, and goods available to the customer. The onus of detecting the need and communicate it to the service provider at the appropriate time is necessary for proper service. The service providers actions are based on the medical protocols to be followed and the availability of resources with them. If the service is medically urgent and the ambulances are available, the services would be dispatched. Depending on the situation, ambulance services employ two operating sub-types, "stay-and-play," i.e., do all possible first-aid at the site; and "scoop-and-run," i.e., transport the patient for further help as fast as possible. Emergency departments in hospitals are specialized facilities designed for emergency cases (Spaite, Criss, Valenzuela, & Guisto, 1995).

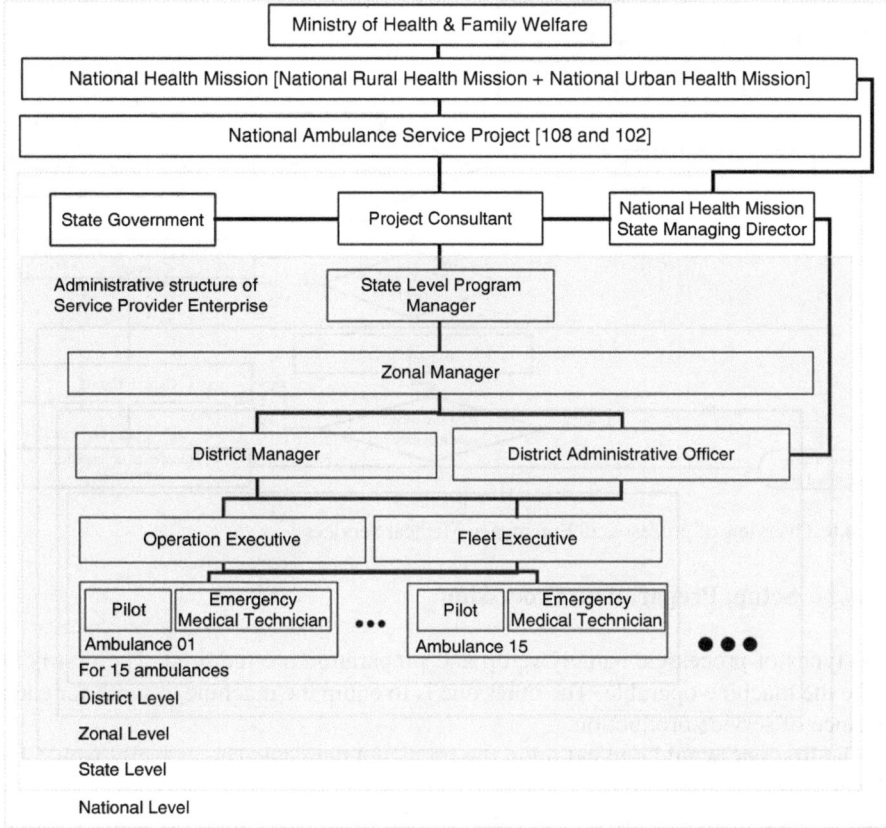

Fig. 4.5 Administrative structure of emergency ambulances services projects in India

As the EMS machine is sensitive to the time, the internal working of the machine needs to be streamlined. A delay due to the working of a component or its communication with other components would affect the performance. A delay in the customer action also may impact the outcome negatively.

4.2.6.1   Process Elements

In the earlier stages of EMS, Sense and Reach, the SM processes the data/information of the customer. The quality of operations depends on the quality of data provided. Later, the SM operates on the property of the customer, i.e., The part of the patient to be treated and subsequently the person would be transported. Apart from addressing the "health" dimension, the patient's mental and psychological factors ("help") have to be addressed all along appropriately to reduce the pain and raise the gains. Thus in this service machine, the customer flows as data, property, and person.

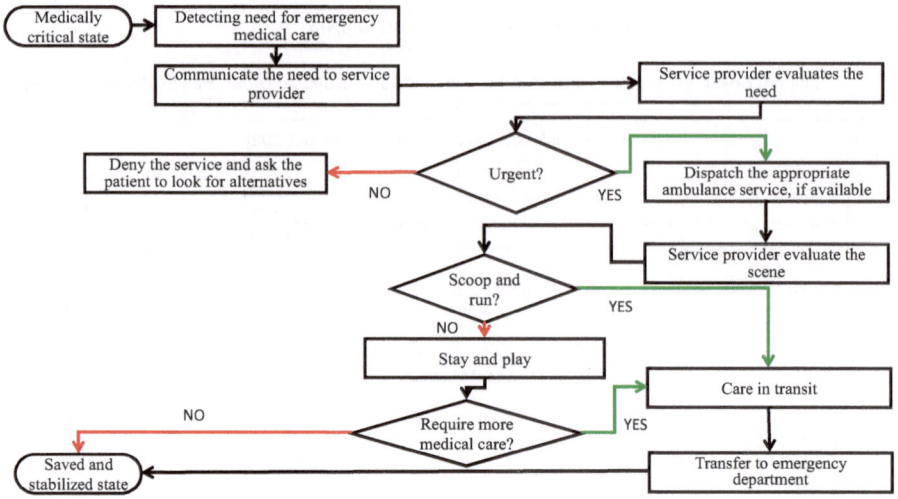

Fig. 4.6 Overview of processes in Emergency Medical Services

4.2.6.2 Setup, Preparation, Processing

Two types of processes, namely setup and preparation are required. The first is to make the machine operable. The other one is to equip the machine to work for each instance of service production.

The first one would depend upon the predicted demand. The capacity would be planned, deploy the resources appropriately and schedule the working. Later for each service instance, there is a need for real-time setup and preparation which would help in preventing the service production from interference from earlier or subsequent service instances. It is critical in the open production system.

As we discussed, there are three types of processing is required in service production as the customer flows as data, property, and person. The processing steps need to handle the flow units meticulously as they are value creation processes and it may result in zero or negative value.

In EMS, it is required to predict the spatio-temporal demand, plan the capacity, and schedule and roster the resources appropriately to set up the system to work. The number, type, and location of the ambulances to be deployed, the type and number of the people to be recruited and their schedule and roster depends on the demand. For each service, it is required to understand the demand type and need to reconfigure the system to feed the demand. As part of setup and preparation for each service instance, the provider needs input from the patient. Based on the information, the provider could understand the situation and configure the services. For example, some of the services may need support from the police, fire or both. Similarly, depends upon the demand the availability of appropriate ambulance has to be evaluated and need to be allocated. These processes are necessary, but they may not result in the state of the patient. There are some processes for which the patient's state changed from uninformed to informed, inaccessible to accessible, and medically critical to treated.

4.2.6.3 Process Layers

A service blueprint of ambulance EMS is presented in Fig. 4.7. Time flows from the top down. Demand is formed in the column "Customer Actions." Supply-side activities are depicted in the column "Backstage activities." Demand and supply meet in the center column "Front stage actions." Here the initial phases are conducted to co-create a design for a possible service instance. This initial phases of the front stage actions can be seen as an act of negotiation of meaning, or sense-making (Weick, 1995). The parties struggle to reach an agreement whether an ambulance should be dispatched or not. Further, the specifics of the situation, such as the number and conditions of victims and the exact location need to be sorted out. In situations where there are more than one feasible EDs, it must be decided where the patient will be taken.

Once the situation has been defined and triage has produced a go –decision, Step 8 is activated, and the Service Machine springs into action. Allocating, dispatching, routing, and directing and ambulance are technicalities that need not be discussed in detail here.

Once the ambulance has delivered the patient to the ED, the field-based EMS has done its job, and the facility-based ED takes over. As long as the patient case is deemed to be critically urgent, activities follow the Emergency DSO (Lilrank, Groop, & Malmström, 2010). It can be envisioned as a demand-supply –informed managerial platform onto which various clinical specialties with specific domain knowledge attach. For example, if the patient has been shot or stabbed, it is the job of trauma surgeons; if there are disjointed and damaged shoulders, orthopedics pitch in; if it is a case of premature birth, an obstetrician is called for. After completing the emergency job, the orthopedic may return to the orthopedic ward and continue following the Elective DSO.

The basic value chain of EMS, the things that must happen for an expected outcome to materialize is the following:

1. An adverse event happens.
2. The event is identified as primarily medical.
3. The event is identified as urgent, i.e., the status of the victim is understood.
4. The people or organizations (EMS) with the capability to help are identified.
5. The EMS is contacted with a request.
6. The EMS assesses the situation
7. The EMS decides to dispatch the ambulance or not.
8. Emergency service processes are executed.
9. The patient is handed over to other caregivers for further treatment, sent home after being patched up, or pronounced dead.

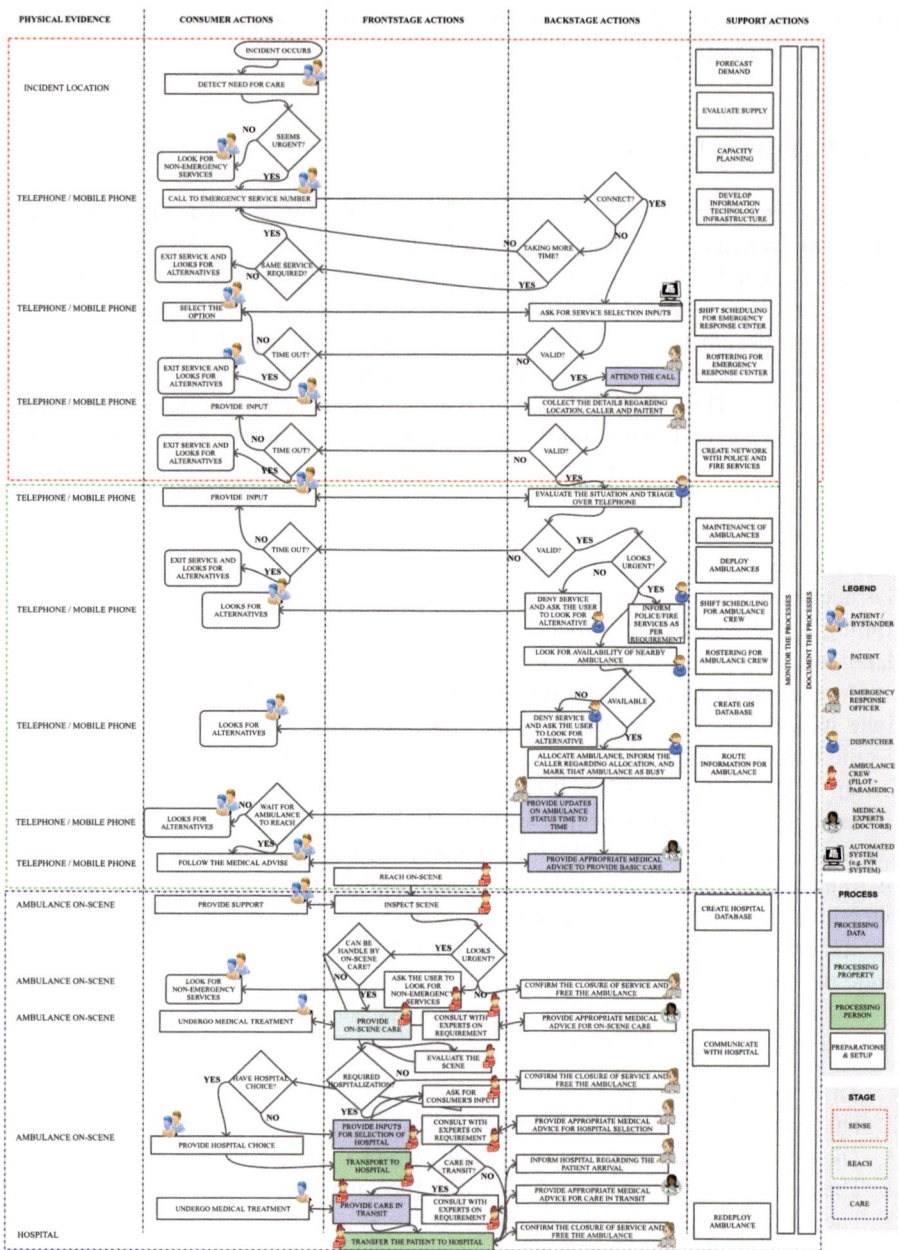

Fig. 4.7 Service blueprint for Emergency Medical Services

Fig. 4.8 Legend
represents the actor and
processes

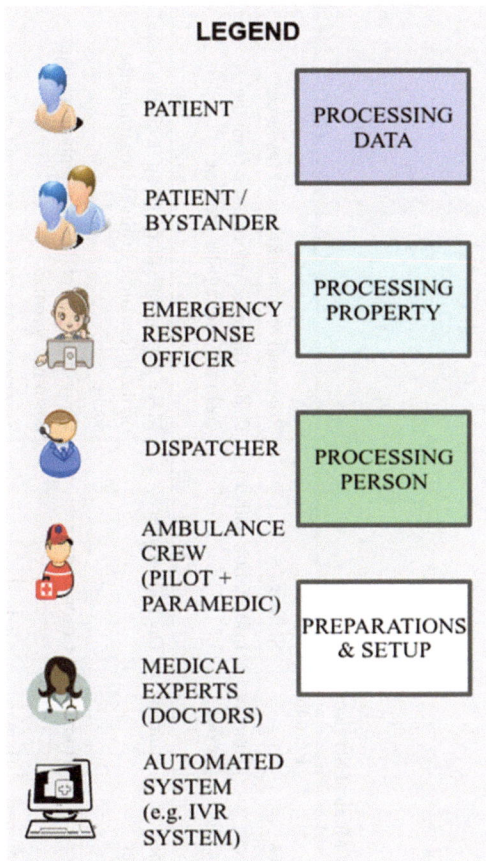

The steps 2–7 imply the planning of a specific service instance, helping patients in one location at one point in time. Steps 8 and 9 are the execution of a service instance, a cycle of emergency care based on a customer request. The service blueprint can be segmented into three: Sense (step 2–5), Reach (step 6–7), and Care (step 8–9).

The actors involved in the EMS service production are patient, a bystander for the patient, emergency response officer, dispatcher, ambulance crew, medical experts, and automated systems (e.g., IVR system). These actors communicate each other and make decisions based on the information available. There are different types of actions carried out by the service providers such as setup, preparation, and processing as discussed earlier. The actors responsible for the action and type of actions are represented in Fig. 4.8 and the role and responsibilities of the actors and the resource group is explained in Table 4.4.

The basic architectural feature of EMS is that step 1 happens randomly in time and space. On a macro-level, emergencies, such as traffic accidents happen with surprising regularity. The number of, say, traffic deaths in a country does not vary

Table 4.4 Roles and responsibilities of resources in Emergency Medical Services

Resource group	Roles and responsibilities of resource group	Resources	Resource type	Skillset	Roles and responsibilities of resources
Call center	Sense the need, dispatch ambulance, Co-ordinate and support the resource on ground	Emergency Response Officer	Human/device	Information gathering	Gather the basic information from the customer and register the case
		Ambulance dispatcher	Human	Decision maker	Make appropriate dispatching decisions based on the information available regarding the scene
		Manager	Human	Co-ordination	Co-ordinate the call center, ambulance, and customer activities
		Medical Experts	Human	Medical	Support the ambulance crew and customer whenever required
Ambulance	Reach the scene, provide on-scene care, care in transit and transfer to definitive care	Ambulance vehicle	Device		
		Pilot (Driver)	Human	Driving, Emergency Medical Technician	Drive the ambulance fast and safely. Provide support for paramedic/nurse/doctor during on-scene care and transfer
		Paramedic/Doctor/Nurse	Human	Medical	Provide medical care to the patients. Help the patients to make decision regarding selection of hospitals and others

more than a few percentages from year-to-year, unless there is a significant mega-trend due to change in road infrastructure and safety. However, on the micro-level, that is the major concern of a single EMS operator, demand is random. There may be periods when nothing happens, followed by periods when everything seems to happen at once. From this follows some central design principles. EMS must have their capacity on standby, ready to take action when needed. Capacity must be reserved based on estimates and averages; from which follows that at times it is idle, at times overburdened. Therefore capacity utilization rate is not a relevant performance measure for EMS. In fact, the less it is needed, the better for society. An EMS cannot meaningfully be financed on a fee-for-service basis.

Another unavoidable uncertainty in EMS follows from steps 2 to 6. The people who attempt to activate the Service Machine may not have a clear understanding of the situation and do not know what to do. Coming back to the bicycle (service machine), somebody wants to ride, but doesn't know how to do it nor where to go. EMS customers typically perceive their need as urgent, and they want fast service. The perceived needs may not be real, as defined by the EMS and its governmental principals. Dispatching an ambulance, or activating an emergency team is costly. Moreover, it commits a resource to a mission, making that resource not available for other requests that may be more urgent.

Sense
In step 2–5, the customer informed the service provider regarding the requirement as shown in Fig. 4.9. Thus it allows the provider to sense the requirement. The customer detects the need for care and contacts the appropriate emergency medical services. The primary step in this phase is to attend the call and register the medical complaint. The acknowledgment from the service provider would help the customer to place the order correctly.

Reach
Therefore step 6, assessment of the situation, also known as triage, is an essential part of EMS architecture. The triage process influences the priority and order of the service offerings (Lidal, Holte, & Vist, 2013). The triage is a crucial process that mixes and matches the demand and supply. It is also a gatekeeper that protects service resources from unwarranted demand (Fry & Stainton, 2005). The triage is done in two basic situations, by the dispatcher receiving a phone call; and by EMS personnel who directly interacts with the patient, either at the site or the ED reception.

The dispatcher must assess the situation based on what the caller says. People under stress do not necessarily articulate well. Two-way communication, interaction, and interpretation of the situation are the most important components in this process. To facilitate this, several types of protocols, assessment guidelines and decision-trees have been developed. For example, in the US authorities have designed a process with three phases (National Highway Traffic Safety Administration (DOT) & Health Resources and Services Administration (DHHS/PHS), 1996).

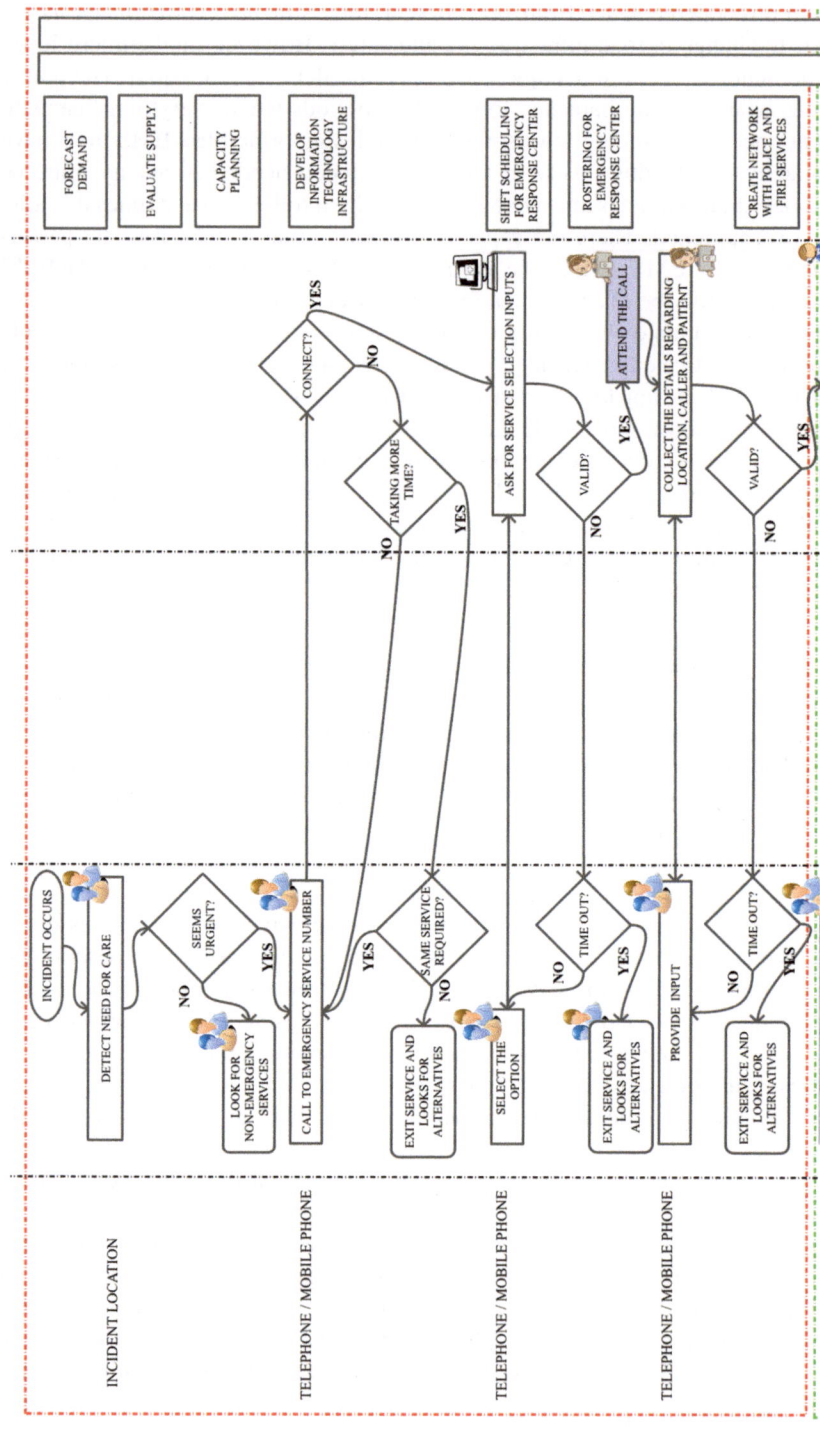

Fig. 4.9 Sense stage in Emergency Medical Services

First comes the "Initial survey" with the questions: where, what, how, who, and when. The order of the questions is important. An answer to the first question will indicate location so that the dispatcher can reach back to the customer back in some way even if there are network issues during the call.

Second, there is a "Chief complaint protocol" by which callers are interrogated to get the details of type and severity of the emergency. The protocol lists 32 chief medical complaints, each requiring different arrangements. This protocol enables the dispatcher to understand the medical need and priority.

Third, a "Scripted medical protocol" is followed to provide instructions to the caller so that the caller can apply life-saving treatment to the victim before the arrival of ambulances.

At step 6, the value is co-created by the dispatcher and the caller. The dispatcher needs to understand the need and respond appropriately. The caller should co-operate and provide the required information. The accuracy of the information affects the service offering and its quality.

The dispatchers are constrained by limited knowledge of the situation; the caller by his or her ability to communicate clearly. If the patient is not responding to the medical protocols, the dispatcher has to take decisions under uncertainty.

When the patient and the EMS medical staff meet in person, a clinical triage protocol is executed. Upon inspection, it classifies the case into urgency types, ranging from immediate action to no action (Fry & Burr, 2002).

The ambulance would be dispatched to the location where the customer reported as per the requirement. The ambulance reaches the location and provides access to immediate medical care as shown in Fig. 4.10.

Based on the triage, the order would be accepted/rejected, and it would be acknowledged by the customer. The ambulance would reach the customer, and the services will be accessible to the customer.

Care

In steps 8 and 9, based on the situation on the scene, the medical care for the patient will be provided. If the patient could be saved and stabilized using the limited resources available with an ambulance, the on-scene care would be provided. If the patient requires further care or the resources available in the ambulance is not enough to provide the sufficient care, the patient would be transferred to the definitive care at the emergency department in hospitals. This is a stage where the state of patient transformed from medically critical to saved and stabilized state as shown in Fig. 4.11.

The resources of the machine mentioned correspond to the skill set and knowledge owned by the machine. For example, the call center as a resource corresponds to the knowledge and skill sets required manage and run a call center, such as organizational, behavioral, and technology. Similarly, ambulance corresponds to the skill set of the pilot (driver) of the ambulance.

In the initial stage, the data associated with a patient and the caller is processed. The information technology-based mechanisms are primarily used in this stages.

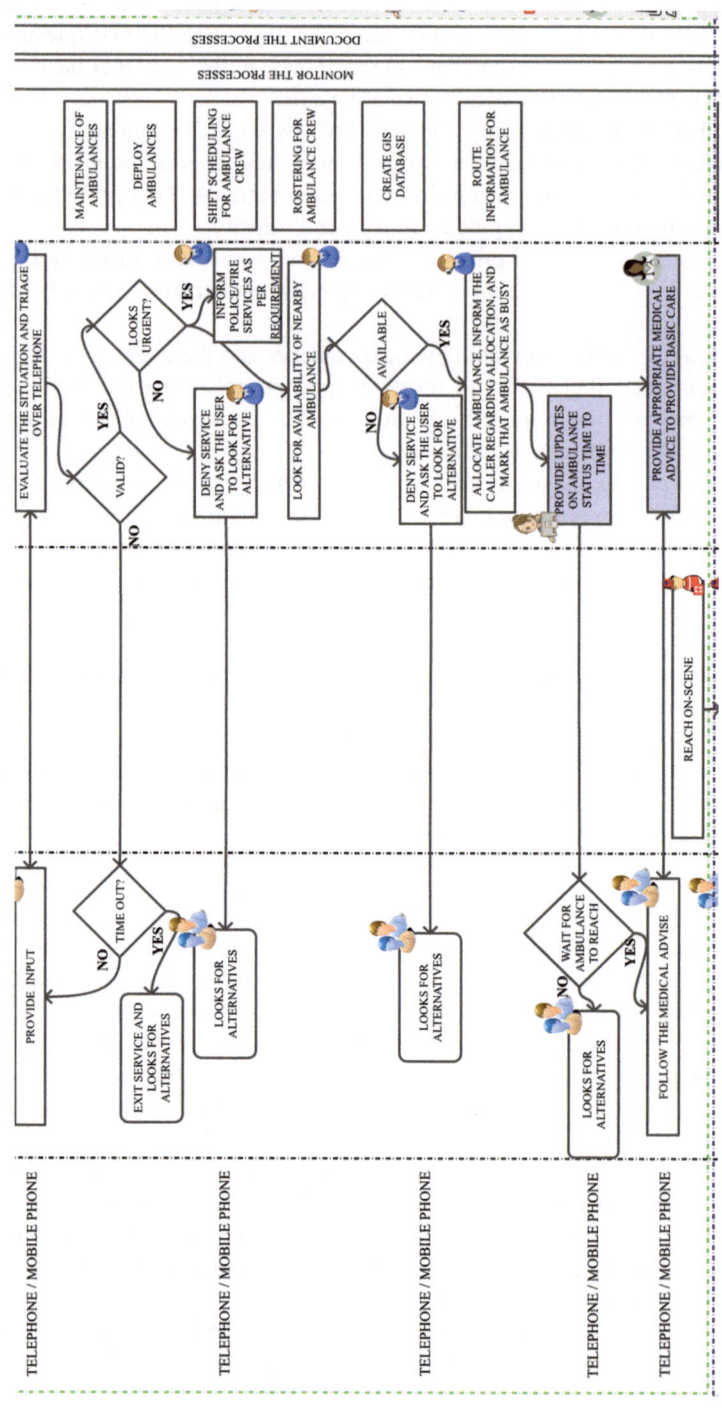

Fig. 4.10 Reach stage in Emergency Medical Services

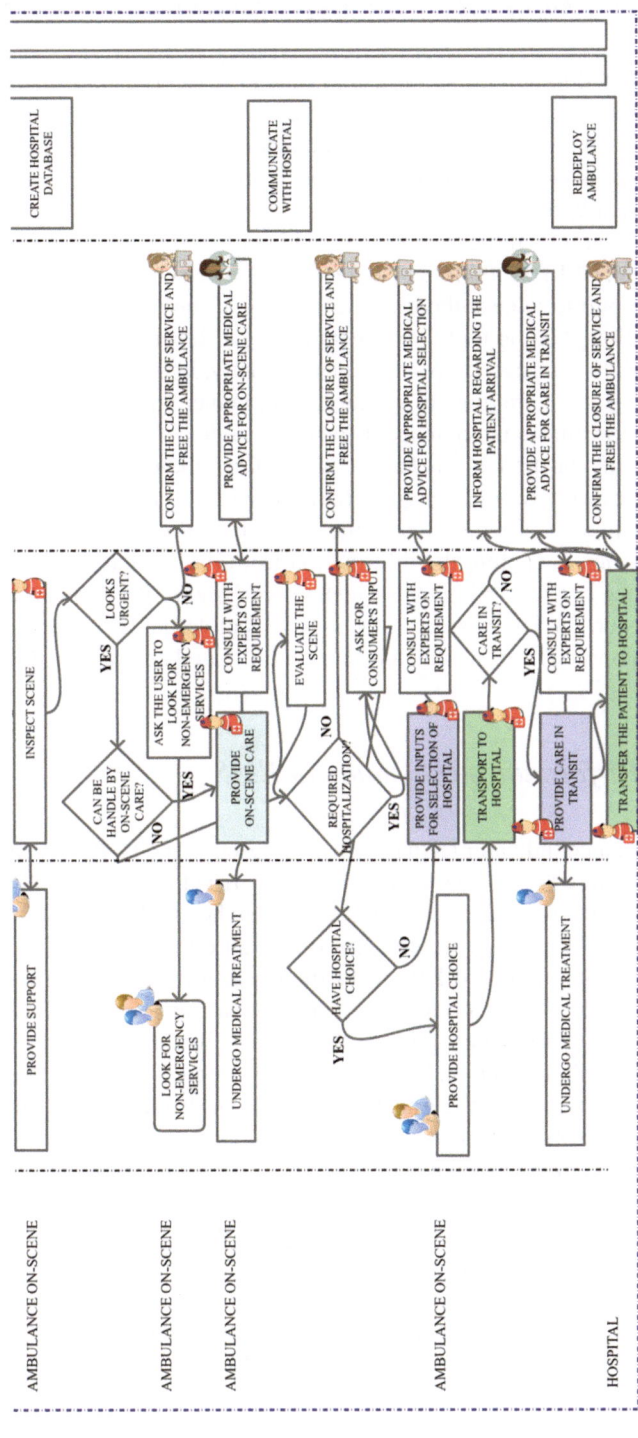

Fig. 4.11 Care stage in Emergency Medical Services

4.2.7 Control, Adjustment, and Improvement

The SM is expected to deliver the expected state change. The state change may be attained but with some variations. The machine's some part is responsible for improving the accuracy and reliability of a machine using its performance indicators. Based on the performance indicators, the appropriate processes would be controlled, adjusted and managed.

In EMS, for example, if the ambulance could not travel as fast as expected due to the traffic, a good EMS would initiate a module to interface with police, traffic personnel, and the public to ease the travel. Similarly, if the time required at requirement identification stage is high, proper training and other supportive mechanism come into action to improve the performance. Usually, in EMS, the control, adjustment, and management are done in a centralized manner with Emergency response center as a hub to control and manage its working nodes.

The benchmark for the EMS is based on the clinical knowledge. Every medical complaint has respective time windows (normally represented as the golden hour and platinum minutes) in which the chances of survival of patients are high. The control system tries to capture the information from the ground the adjust the system to provide the service within the stipulated time frame. The improvement in the system depends on the knowledge of the user to operate the EMS service machine and how the components of the machine are co-ordinated and operated.

4.2.8 Key Performance Indicators in EMS

The key performance indicator (KPI) extensively used in the EMS is response time.

The response time is the time interval between arrival of the call and the time at which the ambulance reach the scene to provide pre-hospital care. This shows how fast the patient could get the services. But this is an output based KPI. There is a requirement for outcome-based KPI. Thus the researchers and practitioners working on another KPI called survival rate. The survival rate signifies how good the service production to save and stabilize the patient. But the issue is that the real survival does not only depend on EMS SM, it also depends on the ED. Hence, normally the state of a patient at the transfer of the patient from EMS to ED is considered for survival rate evaluation.

The time components at each level would affect the output time for services. The major events in EMS and time performance indicators are depicted in Fig. 4.12 (Al-Shaqsi, 2010). This KPI is additive and higher the time component lesser the performance. The Emergency response officer is supposed to attend the call within three rings. Once the call is connected and if the ambulance dispatch is decided then the ambulance should start moving. The call to wheel rolling time is important. If the ambulance pilot and paramedic require more time to start the vehicle, it may affect the performance. The coordination of EMS with other service machines such

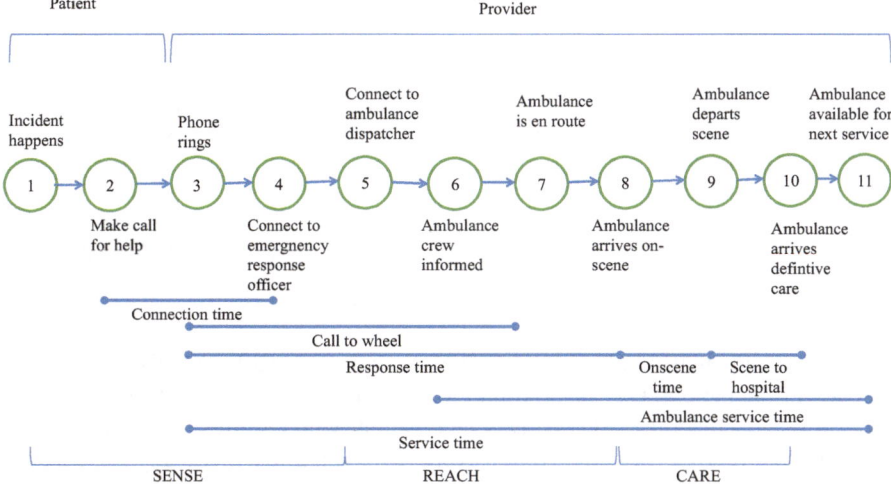

Fig. 4.12 Events and time performance indicators

as traffic police, police, and fire is important in reaching the scene. Once it is decided to transfer the patient to the hospital, then coordination with the customer, traffic, and hospital is also important. Usually, the EMS SM respects the customer's decision in selection of hospitals. Hence it is important to provide necessary information to the customer to make fast and right decisions. The coordination with hospitals/hospital network in important as there are chances of unavailability of resources at EDs of hospitals.

As the on-scene care follows standard medical protocols supplemented with the usage of artifacts available in an ambulance, the on-scene for each complaint should be completed within stipulated time according to the medical complaint. It is important to see the time elapsed at on-scene and total productive time spend there.

4.3   Emergency Department

Emergency Department (ED) is designed to take care of urgent and life-threatening conditions for the patients who require immediate action to save and stabilize their lives. Besides, ED is taking care of less severe urgent cases that may not threaten the life but require urgent examinations and treatment decisions like severe stomach pain or back pain. Thus, ED has two major patient segments (1) patients with life-threatening severe conditions and (2) patients requiring urgent assessment, care or care guidance from a quality of life-perspective.

Once the requirement for urgent care is detected, patients are either they are transported to an emergency department (ED) as a service (EMS), or they transport themselves in less severe cases (walk-in). Once the incident is reported, the provider has to evaluate the situation and respond to the request appropriately within a given

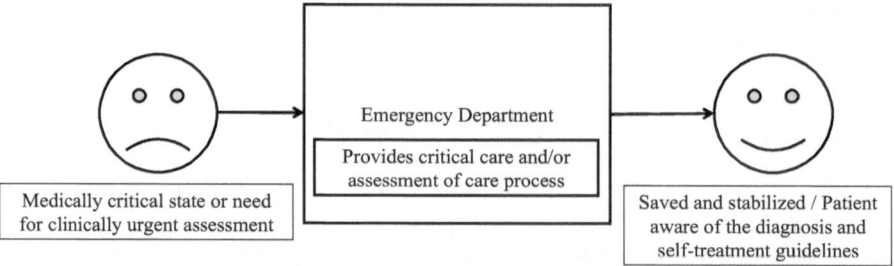

Fig. 4.13 Service machine for Emergency Department

time-frame and an available resource. The evaluation (triage) typically includes the urgency of the patient, the reason for arrival and the decision for care process.

As a service machine, ED operates on a patient who requires immediate care to save and stabilize her life. Thus, it transforms a patient into a medically critical state to a saved and stabilized state using a production function that comprises technologies, such as resuscitation, bandages, tourniquets, and painkillers as shown in Fig. 4.13. The alternative state change is the awareness of the patient concerning the disease or wound causing health problems in case of less severe cases that may not require active care procedures. In those cases, the patient receives the guidance for self-treatment.

To the ED crew, the customer is a patient in person (with some damaged bodily and mental properties), to the diagnostic units (lab, X-ray, etc.) the customer is a case represented by data.

4.4 The Service Machine Template for Emergency Department

In ED the customer is best described as a consumer or a user. Since ED in many countries is a government or charity -funded service, it does not include bargaining between buyer and seller. A patient is, per definition, a person who has requested the services of Healthcare. Thus the ED 'customer' is a patient who arrives at ED and requires immediate assessment or care to stabilize her condition and save her life. The service machine executes the job to be done for the patient. The service template for ED is shown in Fig. 4.14.

4.4.1 Job to Be Done

In ED, the patient comes up with the job to be done (what and how), avoidance of immediate death (mortality) or permanent injury (morbidity). In the first customer segment (save and stabilize) an immediate medical intervention is required to save

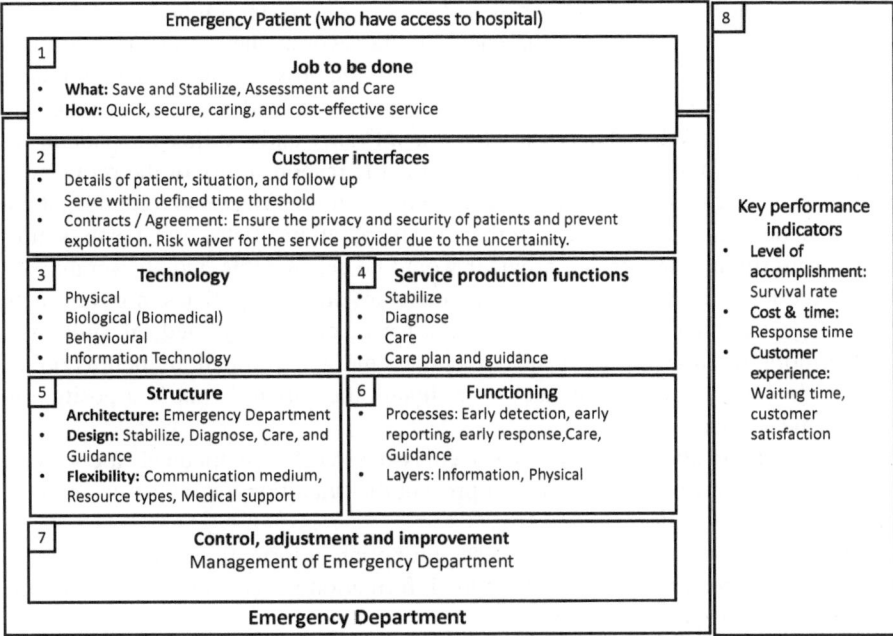

Fig. 4.14 Service machine template for Emergency Department

the life or to avoid morbidity. In the second customer segment (urgent assessment and care) required examinations are needed urgently to find out the diagnosis to assess the required care procedures. The patient comes up with the requirement of gains and acknowledges that there are some inevitable pains from and during the services. The patient needs to go through events that involve uncertainty (e.g., waiting for the doctor or examinations) and mental and physiological stress.

4.4.2   Customer Interfaces

Similar to EMS, EDs are designed for rapid response. In this case, also, the cost involved is high as there is a need for slack specialized resources and needs to ensure availability. Again, the customer could either walk-in or by ambulance. The gatekeeping mechanism is somewhat easy if the patient comes by ambulance. Assessing the urgency of walk-in patients is difficult.

As patient arrives the ED, triage made by nurse or doctor evaluates the priority and reason for arrival using standardized protocols. Based on the triage, the customer is segmented, and the care process decided. The save and stabilize -patients are transferred immediately to a specific room for immediate care. The other patients (urgent assessment and care) have ED-specific care processes like fast-track, traumas, and others.

The patients come to ED, either by ambulance or walk-in, implicitly informs the service provider that there is an urgency. In other words, the customer agrees that the state of the customer is critical and actions of the provider may or may not have an impact on recovery. By accepting the customer, the providers accept that they are ready to provide quick and necessary treatment to save and stabilize the patients. In case of non-availability of definitive care, it is the responsibility of the provider to inform the customer regarding the unavailability of the services and refer the patient to nearest definitive care services. If there is an operational delay and it causes mortality or morbidity, the service provider would be accountable and answerable. As part of reducing the risks over providers, there are different policies in government to manage the medico-legal issues may arise in emergency medical situations. Furthermore, to ensure medical services in an emergency situation, there is policy to provide emergency medical care even though the patient is not in a position to pay for the services.

The policies to alleviate the medico-legal issues in different medical emergencies is important as those issues would not prevent the practitioners to operate and provide services.

The patient is expected to provide the appropriate information to the provider to understand the requirement. Based on the information provided by the patient, the care process is decided.

As these services are information intensive, both parties agree to share appropriate and actual information for mutual benefit. The providers could use the information only within the legal and ethical boundaries of a country.

4.4.3   Technology

Different technologies such as information, diagnostic, medical, and behavioral technologies are used in ED. The information technology is important for communication between the provider and customer, and internal communication of provider. The diagnostic technologies may include lab, x-ray, CT and other tests. The medical and behavioral technologies are required to ensure appropriate medical services and patient's mental and physiological support.

Table 4.5 illustrates how the supply side has applied new technologies to meet customer requirements better. The urgency –imperative calls for speed in all steps.

4.4.4   Service Production Function

In ED, the core service is the stabilize, diagnose, care and guidance. The service engine makes use of technologies such as IT, diagnostic biomedical, and behavioral technologies to operate on the different customer flow units.

The service production in ED include four primary state changes, and these state changes are brought by the service production functions: Stabilize, Diagnose, Care,

Table 4.5 Technology available for the job to be done in Emergency Department

Customer requirements/ demand	Description on the value creation	Technology
Want to get in touch with an actor who can provide advice and help in an emergency, and communicate the urgency and the details of the situation and its location	Enhance access, improve communication, and support co-design of the service instance.	Introduction of nation-wide or state-wide information of the guidelines for the patients in urgent needs and the availability of ED services. (behavioral)
		Easy access to Emergency Department of the hospital (behavioral and structural)
		Algorithms and decision trees for quick and accurate assessment of urgency. (bio-medical and information technology)
Want to get quick access to medical care	Depends on the urgency and medical condition various technologies are evolved. The logic for service is decided based on the interaction between the customer and provider. These services are offered by the providers to enable faster and appropriate access.	Save and stabilize:
		Doctors, nurses and specialized equipment, such as monitors, trauma kits, defibrillators, etc. (bio-medical and information technology)
		Assessment and care:
		Doctors, nurses, and specialized equipment, such as lab, radiological, care kits etc. (bio-medical and information technology)

Guidance. In the Stabilize service production function, the customer's medical condition is stabilized to prevent morbidity or mortality. In the Diagnose phase, two cognitive transformations happen. The customer would transition from uninformed to informed state, knowing the reason for symptoms. The provider similarly transfers from an uninformed to an informed state; the care procedures can be decided based on diagnosis. Care is the production function of giving care. Guidance is for patients who do not require care procedures in the ED but may need services in other service units or self-treatment. The resources, technologies, and the state change involved in service production function for ED are depicted in Table 4.6.

4.4.5   Structure in ED

In a typical ED, the customer as a user and customer as a payer is different. Usually, the customer as payer is the government which outsources the work to a service provider or allocate the work to the department of government. The customer as a user could be the patients him/herself or a person who arrives at ED and uses ED

Table 4.6 Resources and technologies in service production function for Emergency Department

Production function	Machine resources involved	Technologies	State change in machine	Role of customer	State change in customer
Stabilize	Doctor, nurses, medical equipment	Information, bio-medical behavioral, diagnostic	Patient stabilized for further examination and procedures	Customer is typically passive in this phase	Stabilized medical condition
Diagnose	Doctor, nurses, diagnostic equipment	Information, behavioral, diagnostic	Informed	Understanding the situation and report the same	Informed
Care	Doctor, nurses, medical equipment	Information, behavioral, bio-medical,	Processing	Possibly collaborate with the professional and undergo treatment	Healing process is initiated
Guidance	Doctor, nurses	Information, behavioral	Ready	Collaborate with the professional	Patient is aware of the required further actions

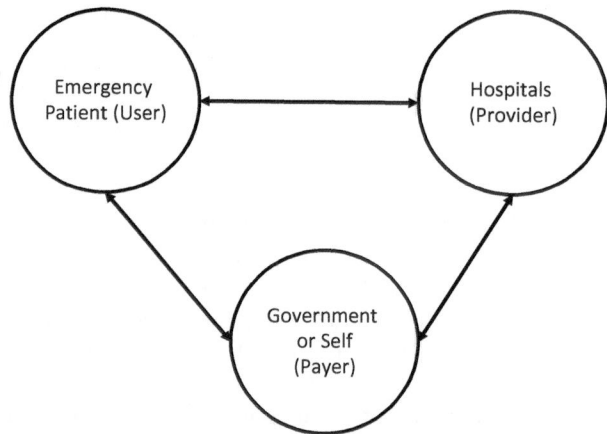

Fig. 4.15 Agents in ED system

services. The payer would not involve in the process level operations while the user plays a role in co-creating the value during each service. The primary inputs from the user are information and the user itself. Figure 4.15 depicts the three agents in a usual ED system where the system is funded by Government. In this case, the Government would act as the principal and service provider would act as the agent for Government to provide the services to its citizens. The government could monitor the working of ED as in modern ED systems have most of the activities are captured using Information Technology (IT) systems. The systems where ED

Table 4.7 Work allocation, interaction, and verifiability in Emergency Department

Phase	Process	Work	Interaction	Verifiable consumer actions	Verifiable provider actions	Service production functions
Detection	Standard	Consumer	Low	No	NA	Stabilize
Stabilize	Standard	Provider-consumer	High	No	Yes	
Diagnose	Standard	Provider-consumer	High	Yes	Yes	Diagnose
Care	Standard	Provider	Low/ medium	Yes	Yes	Care
Guidance	Standard	Provider-consumer	High	Yes	Yes	Guidance

provider is paid by the patient directly would have a different structure. In those cases, the payer would be patient him/herself or insurance provider. In every case, the underlying relations between the agents remain more or less same.

The core service functions are to save and stabilize and assessment and care. The procedures followed during the services are governed by standard medical protocols. The user needs to play the role of provider of information and input to the medical procedure as him/herself. The involvement of user and service provider in ED varies depending on the medical condition of the user and the phase of ED process.

The work allocation, interaction, and verifiability, in the ED service are shown in Table 4.7. We have four core service production functions in ED: Stabilize, Diagnose, Care and Guidance. In stage Stabilize, the provider provides the appropriate care to stabilize the medical condition. In the other phases, the collaboration varies depending on the medical condition of the patient.

4.4.6 Functioning and Processes

The overview of processes involved in transforming a medically critical patient or patient in urgent need of assessment and care to a saved and stabilized state is shown in Fig. 4.16. The initiation of the services depends upon the knowledge, skills, and goods available to the customer. The onus of detecting the need and communicate it to the service provider at the appropriate time is necessary for proper service. The service providers actions are based on the medical protocols to be followed and the availability of resources with them. If the service is medically urgent, the patient will be immediately transferred for immediate care. In other cases, depending on the situation, ED services employ three operating sub-types: "diagnose", i.e., the symptoms don't need further actions based on diagnosis; "care", i.e., the care needed to initiate the healing process can be provided in the ED; and "guidance", the care actions are in other service units or self-treatment. Emergency departments in hospitals are specialized facilities designed for emergency cases.

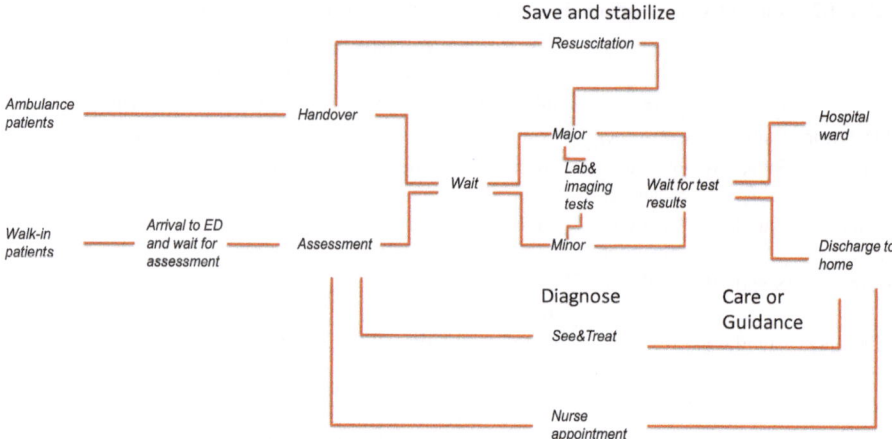

Fig. 4.16 Overview of processes in Emergency Department

As the ED machine is sensitive to the time, the internal working of the machine needs to be streamlined. A delay due to the working of a component or its communication with other components would affect the performance. It is applicable for the customer actions also.

4.4.6.1 Process Elements

In the earlier stages of ED (Triage) the SM processes the data/information of the customer. The quality of operations depends on the quality of data provided. Later, the SM operates both on the property and the information of the customer, i.e., The part of the patient to be treated and subsequently the information is analyzed in the diagnostic process. Apart from addressing the "health" dimension, all along the patient's mental and psychological factors ("help") have to the addressed appropriately to reduce the pain and raise the gains. Thus in this service machine, the customer flows as data, property, and person.

4.4.6.2 Setup, Preparation, Processing

There are two types of setup and preparation required. The first is to make the machine operable. The other one is to equip the machine to work for each instance of service production.

The first one would depend upon the predicted demand. The capacity would be planned, deploy the resources appropriately and schedule the working. Later for each service instance, there is a need for real-time setup and preparation which would help in preventing the service production from interference from earlier or subsequent service instances. It is very important in the open production system.

As we discussed, there are three types of processing in service production as the customer flows as data, property, and person. The processing steps need to handle the flow units meticulously as they are value creation processes and it may result in zero or negative value.

In ED, it is required to predict the spatiotemporal demand, plan the capacity, and schedule and roster the resources appropriately to set up the system to work. The number, type, and schedule of professionals with different specialties to be recruited, the type and number of the diagnostic and treatment facilities and equipment have to be planned. For each service, it is required to understand the demand type and need to reconfigure the system to feed the demand. As part of setup and preparation for each service instance, the provider needs input from the patient. Based on the information, the provider could understand the situation and configure the services. For example, some of the services may need resources from other units of the hospital. These processes are necessary, but they may not result in the state of the patient.

4.4.6.3   Process Layers

A service blueprint of ambulance ED is presented in Fig. 4.17. Time flows from the top down. Demand is formed in the column "Customer Actions." Supply-side activities are depicted in the column "Backstage activities." Demand and supply meet in the center column "Front stage actions."

Once the situation has been defined and triage has produced a decision, The Service Machine springs into action.

Once the ambulance has delivered the patient to the ED, the field-based EMS has done its job, and the facility-based ED takes over. As long as the patient case is deemed to be critically urgent, activities follow the Emergency DSO (Lillrank et al. 2010). It can be envisioned as a demand-supply –informed managerial platform onto which various clinical specialties with specific domain knowledge attach. For example, if the patient has been shot or stabbed, it is the job of trauma surgeons; if there are disjointed and damaged shoulders, orthopedics pitch in; if it is a case of premature birth, an obstetrician is called for. After completing the emergency job, the orthopedic may return to the orthopedic ward and continue following the Elective DSO.

The basic value chain of ED, the things that must happen for an expected outcome to materialize is the following:

1. An adverse event happens.
2. The event is identified as primarily medical.
3. The event is identified as urgent, i.e., the status of the victim is understood.
4. The people or organizations (ED) with the capability to help are identified.
5. The ED is contacted with a request.
6. The ED assesses the situation
7. The ED decides to care the patient or not.
8. The patient is stabilized if required

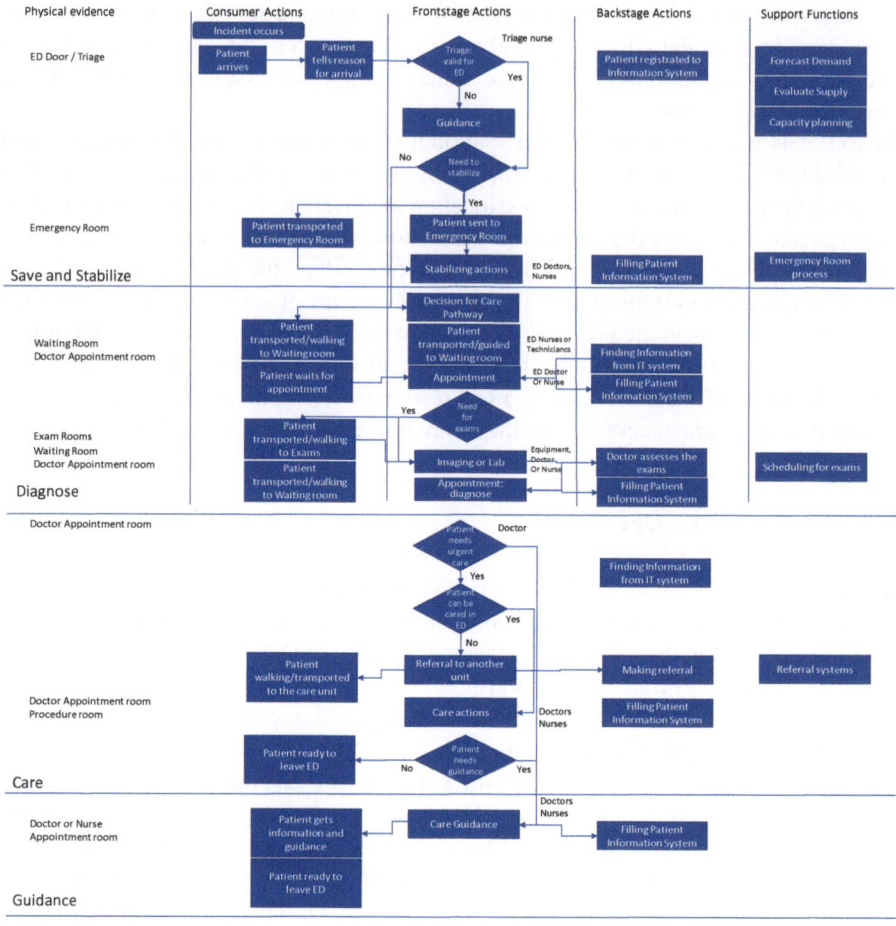

Fig. 4.17 Service blueprint for ED

9. The patient is diagnosed
10. The urgent care or guidance is served
11. The patient is handed over to other caregivers for further treatment, sent home
     after being patched up, or pronounced dead.

The steps 2–7 imply the planning of a specific service instance, helping patients
in one location at one point in time. Steps 8 and 9 are the execution of a service
instance, a cycle of emergency care based on a customer request. The service blueprint can be segmented into *Stabilize, Diagnose, Care* and *Guidance*.

The actors involved in the ED service production are patient, a bystander for the
patient, triage personnel, medical experts, and automated systems (e.g., IVR system). These actors communicate each other and make decisions based on the information available. There are different types of actions carried out by the service

Table 4.8 Roles and responsibilities of resources in Emergency Department

Resource group	Roles and responsibilities of resource group	Resources	Resource type	Skillset	Roles and responsibilities of resources
Triage	Assess the urgency, reason for arrival and care process	Triage secretary	Human	Information gathering	Gather the basic informationfrom the customer and register the case
		Triage nurse	Human	Decision maker	Make appropriate decisions based on the information available regarding the scene
Care line	Stabilizes, diagnoses and provides urgent care and guidance	Rooms	Facilities		
		Doctor Nurse	Human	Emergency medicine	Provide medical care to the patients. Help the patients to make decision regarding selection of hospitals and others
		Lab/X-ray/etc	Device	Diagnostic tests	Diagnostic devices used for diagnosing the patient

providers such as setup, preparation, and, processing as discussed earlier. The roles and responsibilities of resources in Emergency Department are shown in Table 4.8.

The basic architectural feature of ED is that step 1 happens randomly in time and space. On a macro-level, emergencies, such as traffic accidents happen with amazing regularity. The number of, say, traffic deaths in a country does not vary more than a few percentages from year-to-year, unless there is a significant mega-trend due to change in road infrastructure and safety. However, on the micro-level, that is the major concern of a single ED operator, demand is random. There may be periods when nothing happens, followed by periods when everything seems to happen at once. From this follows some central design principles. ED must have their capacity on stand-by, ready to take action when needed. Capacity must be reserved based on estimates and averages; from which follows that at times it is idle, at times overburdened. Therefore capacity utilization rate is not a relevant performance measure for ED. In fact, the less it is needed, the better for society. An ED cannot meaningfully be financed on a fee-for-service –basis.

Another unavoidable uncertainty in ED is that the people who attempt to activate the Service Machine may not have a clear understanding of the situation and do not know what to do. Coming back to the bicycle (service machine), somebody wants to ride, but doesn't know how to do it nor where to go. ED customers typically perceive their need as urgent, and they want fast service. The perceived needs may not be real, as defined by the ED and its governmental principals. The patient commits a resource to a mission, making that resource not available for other requests that may be more urgent.

The customer informed the service provider regarding the requirement. Thus it allows the provider to sense the requirement. The customer detects the need for care and contacts the appropriate emergency medical services. The main process steps in this phase is to attend the call and register the medical complaint. The acknowledgment from the service provider would help the customer to place order properly.

Therefore step 6, assessment of the situation, also known as triage, is an essential part of ED architecture. The triage process influences the priority and order of the service offerings (Lidal, Holte, & Vist, 2013). The triage is a crucial process that mixes and matches the demand and supply. It is also a gatekeeper that protects service resources from unwarranted demand (Fry & Burr, 2002; Fry & Stainton, 2005).

The triage is done in two basic situations, by the dispatcher receiving a phone call; and by ED personnel who directly interacts with the patient at the ED reception.

The dispatcher must assess the situation based on what the caller says. People under stress do not necessarily articulate well. Two-way communication, interaction, and interpretation of the situation are the most important components in this process. To facilitate this, several types of protocols, assessment guidelines and decision-trees have been developed.

Then the value is co-created by the dispatcher and the caller. The dispatcher needs to understand the need and respond appropriately. The caller should co-operate and provide the required information. The accuracy of the information affects the service offering and its quality.

The dispatchers are constrained by limited knowledge of the situation; the caller by his or her ability to communicate clearly. If the patient is not responding to the medical protocols, the dispatcher has to take decisions under uncertainty.

When the patient and the ED medical staff meet in person, a clinical triage protocol is executed. Upon inspection, it classifies the case into urgency types, ranging from immediate action to no action (Fry & Burr, 2002).

Based on the triage, the order would be accepted/rejected, and it would be acknowledged by the customer.

Next, the medical care for the patient will be provided. If the patient requires stabilization, the ED resources available will provide the saving and stabilizing care. Then the patient will be diagnosed using assessments by a medical professional and required diagnostic test. The urgent care and guidance is given in the ED, but if the patient requires further care or the resources available in the ED is not enough to provide the sufficient care, the patient would be transferred to the definitive care at the ward in hospitals. This is a stage where the state of patient transformed from medically critical to saved and stabilized state.

4.4.7　Control, Adjustment, and Improvement

Usually in all hospitals, the emergency cases are treated through a window of service which is different from other DSOs. Emergency Department follows classic triage procedure (color coded) for managing the cases easily. Depending on severity and urgency, the patients will be treated. Typically, a dedicated team of doctors,

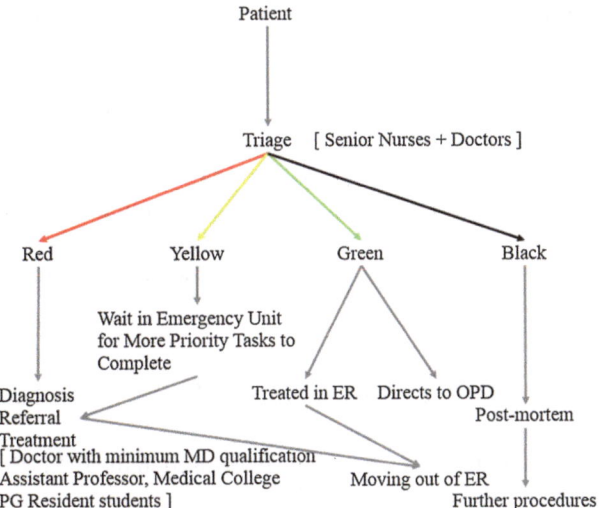

Fig. 4.18 Patient flow in Emergency Department

nurses, and technician takes care of the department. Facilities for basic diagnostics and medicines are available in emergency department itself. The department avail the services from the resources of different other department on demand. Hence most of the hospitals have appropriate facilities to bring the resources quickly on demand. The co-ordination and integration of the resources for diagnosis and treatment attributes to the main issues in emergency care. The delay in the availability of diagnostic results is one of them.

For example, the patient flow in an emergency department at one of the hospitals in India, is as depicted in Fig. 4.18. The emergency room (ER) is designed to save and stabilize the patient and cater their needs up to 6 h. The patients which are identified with color code red will be taken care immediately by the emergency care team (a doctor, assistant professor and post graduate (PG) residents). They can admit the patients to any department directly, if it is clinically required. This reduce the delay due to consultation. The availability of basic diagnostic facilities and medicines at emergency care make the care fast and efficient.

There is no separate emergency care service channels available. But the beds in emergency room is dedicated to specific cases – especially second and sixth bed. The second bed will be kept free every time to attend any emergency case. Once that case is taken care (i.e. once it saved and stabilized) the patient will be moved to another bed or out of ER. Similarly the sixth bed is dedicated for the service of neonatal baby issues.

The emergency care unit is equipped to call the doctors from respective departments in the hospital on demand. Furthermore it is equipped to cater the demand of mass casualty up to 50 patients at any time. The basic necessary medicines and accessories are kept separately for mass casualty. It will be treated as Code Yellow and the assigned doctors from all departments will be alerted once it is reported. The waiting area will be converted into ward and services will be provided as soon as possible.

The emergency department segments its customer and manages them according to their needs. The department needs to have communication facility to bring the help of specialists on demand. Thus the management needs to be dynamic. The decision power of doctors in emergency helps to reduce the delay in operational procedures to an extent. The doctors are permitted to admit the patient to any department if required which will reduce the crowding. The availability of basic lab facilities and medicines in emergency care unit itself make the emergency care faster.

4.4.8   Key Performance Indicators in ED

The key performance indicator (KPI) extensively used in the ED is throughput time, and the sub-measure for that is time-to-doctor.

The throughput time is time interval between arrival or call of the patient and the time at which the ED has performed the urgent care, guidance or diagnosis. Time-to-doctor is defined as a time interval between arrival or call of the patient and the first doctor appointment. These shows how fast the patient could get the services. But this is an output based KPI. There is a requirement for outcome-based KPI. Thus the researchers and practitioners working on another KPI called survival rate. The survival rate signifies how good the service production to save and stabilize the patient. But the issue is that the real survival does not depend on ED only, it also depends on the case-mix and the further services like wards and operating theatre. Another problem is that the survival rate is a very rough measure to evaluate the outcomes of the ED. More detailed patient-reported outcome measures (PROM) are needed to assess the outcomes of the EDs.

4.5   Interfacing Service Machines

To produce a service or to transfer the output of service to another service machine, the service machines need to connect. The interfaces between the service machines are essential in this case. The customer would be at particular state and interface should allow the customer flow as a person, data, property and their combinations. Along with that it also needs to enable the sharing of information, goods, work, and risk.

In case of an emergency, the EMS may need support from the other service machines like police, fire, and so on as it could be a medico-legal problem or incident where fire rescue team has to provide necessary support to make medical care accessible to the patients. Similarly, once the on-scene care is done the patient may need to transfer to another service machine like emergency department (ED) in a hospital. Thus it is necessary to have interfaces with all these types of service

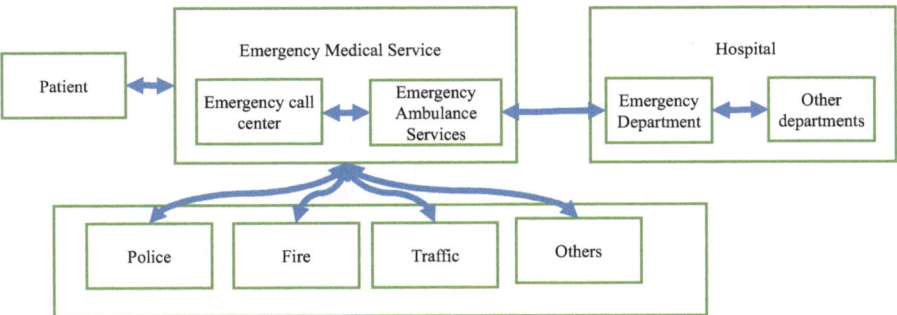

Fig. 4.19 Interconnected services machines in Emergency Medical Services

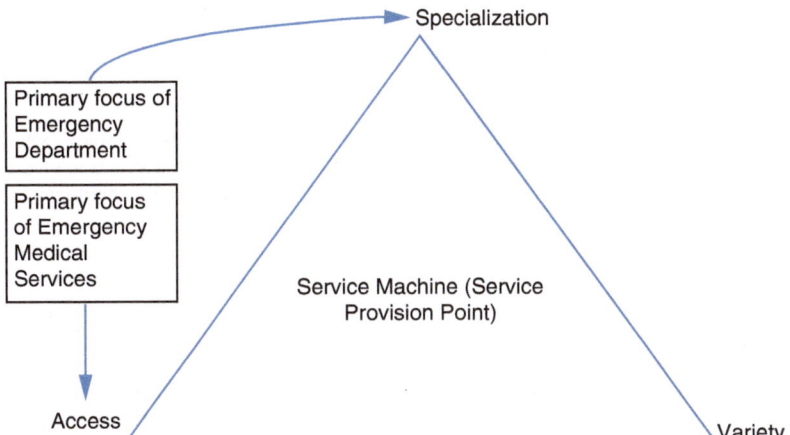

Fig. 4.20 Service provision point triangle

machine for a smooth functioning of EMS. The EMS SM is interconnected to different SM as depicted in Fig. 4.19.

4.6   Discussion

The service machines for EMS and ED have many overlapping components as they are catering similar, but not identical customer segment. EMS is more for increasing the access to the save and stabilize service; the ED is improving the Emergency Medicine specialization to save and stabilize the patient. Thus these two SM have the different focus point on the service provision triangle – access, specialization, and variety as shown in Fig. 4.20. ED also have significant consideration for access to the patient. Both the machines segment themselves for providing the variety of many as specialized and fast services for trauma, cardiac, pregnancy, and others. Thus as the focus of the SM is different appropriately the interfaces, technologies,

Table 4.9 Similarities and dissimilarities in service machines for Emergency Medical Services and Emergency Department

Service machine component	Similarities	Dissimilarities
Customer segment	Need immediate medical care for save and stabilize the patient in both EMS and ED	In case of EMS, the patient needs an intermediate service to access the emergency medical services or need some level of medical support to save life. In case of ED, the patient got access to the emergency medicine specialization. The specialized and advanced definitive care (including surgeries) to save life.
Customer Interface	The customer interface needs to have easy access, ability for two-way communication for understanding care requirements, negotiation, care pathways, and risks.	In EMS, the user communicates with a remote assistant during call and crew directly during on-scene care. In ED, the interactions are primarily direct.
Technologies	Medical, behavioral, and information technology.	EMS needs automobile technology while ED needs specialized diagnostics and medical technology on the top of technologies available in EMS.
Service production function	Sense and care	For ED, the reach is not as applicable for EMS. ED require a focus on another production function advice, to provide necessary advice to the patients to save lives.
Structure	The structure include the patient and medical practitioners with appropriate medical and information technology with time frames.	In EMS, the structure is defined over mobile care unit with minimal and essential care technologies. In ED, the definitive care using sophisticated and advance technologies with time frame.
Processes and functions	Defined healthcare pathways for each diseases.	Call center operations in EMS which is not part of ED. the high level diagnostics and advise is not part of EMS.
Control and management	Co-ordinate and integrate the different technologies within stipulated time frame	On-demand support from different departments of the hospital is used in ED, while in EMS, the services primarily done by the resources.
Key performance indicators	Survival rate is applicable for both the SMs	The most commonly used measure in EMS is the response time, the time it requires to access the care. In ED, usually the throughput time or waiting time is used apart from survival rate.

and service production function changes. This will lead to having a different structure, functioning, control, and management. Lastly, but not the least, the KPIs changes. Thus mostly we talk about the response time as KPI for EMS while survival rate for ED.

Similarities and dissimilarities in service machines for Emergency Medical Services and Emergency Department are shown in Table 4.9. The similarities in these SMs help to integrate them to produce the value. One service machine starts where the other one stops. The main problem in the integration of these two service machine is the level of structure. Although the architecture and design of these machines are clear and primarily follows them, in different locations, these machines have different levels of flexibility. If there is flexibility mismatch, the system integration becomes difficult. In India, we have multiple EMS models which are identical in architecture and design. Due to the flexibility in implementation, each state has different levels of EMS facilities. Similarly, regarding architecture and design, the ED operations also matches. The flexibility allows the SMs to provide the better services in each service instance level. Understanding the state of an EMS SM at each instance and adjusting the ED SM for the same is a difficult task. The recent developments in technologies regarding artificial intelligence, and communication (using the Internet of Things (IoT) sensors), the integration and coordination become easier. The ED SM could predict the state of EMS SM in advance and adjust itself for the input.

Recent development in data capturing (using sensors, IoT), data processing (big data analytics) is vital for a service machine designer. The current data analytics and artificial intelligence are helping the providers to understand the physical and mental state of the patients to a large extent. The communication technology which is becoming faster and faster. The human involved int system is becoming the bottleneck. The aid provides machines to make decisions improving its accuracy and helps the human to make fast informed decisions. But, the medical world is probably the last one to make use of it. The lag in technology adoption, the resistance from the people in power (due to significant information asymmetry and skepticism), and self-service nature (many times the patients are the ones to adopt the technology) are some of the reasons for this. Thus when designing the service machine in healthcare, the designer needs to take care of these reasons. The structure should be able to accommodate these reasons otherwise the technology selected may not work as expected.

The concept of SM is useful in understanding the working of EMS and ED better. It helps in defining the service production functions and assists the providers in selecting appropriate technologies and structuring the same to perform particular processes and functions involved in EMS and ED.

Chapter 5
Summary and Discussion

Abstract Every service production system can be described by the categories explicated by the Service Machine template. This template is like a lens which allows one to see the things which we could not see otherwise. This helps in understanding the service production system better. The development of the Service Machine template is not always a straight forward task. Some of the challenges faced in the development of the Service Machine template are discussed. One of the direct uses of the Service Machine concept is to analyze and decipher the problems in an existing service production system. This concept will help in examining probable reasons of failure of a service production system that otherwise may be difficult to identify. The advances in technologies are providing better tools, but new challenges to the service production system designers. The concept of Service Machine is expected to help the designers, entrepreneurs, and researchers in identifying and overcoming these challenges and making better use of technologies in service innovation and entrepreneurship.

Keywords Online services · Systematic descriptions · System analysis · Service production · Service entrepreneurship

The Service Machine builds on the Universal Machine as Biology builds on Chemistry. In a hierarchical system, such as the world expressed by areas of science ranging from particle physics to the psychology of emotions, a higher level brings new system dynamics that are not entirely explainable by the underlying layers. Never the less, the higher levels cannot violate the basics of the fundamental. Behavioral sciences can't ignore Biology; Biology can't go against the basic rules of Chemistry. From this follows that every service production system can be described by the categories explicated by the Service Machine. Each particular service brings its specific technologies, demands, and customer segments. These can be interpreted and arranged into systematic descriptions.

We have chosen EMS as the case since it is a relatively straightforward Service Machine. The primary benefactor, the patient, may not be the agent that actively participates in the initial steps of designing the service instance and co-creating value in Emergency Medical Services (EMS) and Emergency Department (ED).

Friends, relatives or anonymous bystanders may have to get involved as secondary customers on short notice. Therefore, both the customer and the types of value created are complex. In most of the countries, the payers are governments as the primary value of EMS is that it offers legitimacy to governments. Citizens want the sense of safety and security. Therefore, emphasis on mere economic efficiency calculations can be misleading. This is, in some countries exemplified by the debate on ambulance helicopters. They are expensive to operate. The same money, if spent on rubber-wheeled ambulances saves more lives. Still, people demand helicopters, as the image of a rescuer from the sky offers a compelling sense of safety (Robers, 2011).Given that EMS is usually a public- or third-party financed service, the customer does not pay for it. Therefore, in the specific situation, the customer is in a weak bargaining position. Due to the multiple customers, the customer requirements become complicated, sometimes even counters each other. The payer would like to reduce the cost, while the users would like to have all the facilities and as fast as possible. Thus the contracts in EMS and ED are difficult to lay down.

Many times, the service production is a continuous process. Thus demarcation of system boundaries is complicated. In case of EMS and ED, there is an overlap in the processes. The physical boundary of the service production system – inside and outside of hospital's ED – is considered for defining the boundaries of service machines. Anybody with a smattering of exposure to EMS can easily identify tasks, processes, and participants. It may sound odd to describe the act of stopping a person from bleeding to death as a 'state change to a stable state', or a tourniquet as a 'technology' which is part of the 'production function' that deals with open wounds. The abstract language, however, serves the purpose of analyzing service systems where the Machine is not as clear and visible as in EMS. As an example, think of a clinic that aims at preventing diabetes? What would be the state change and how would it be measured? How would you identify and evaluate the technologies and production functions? The prevention services are difficult as the outcome is nonexistent. Measuring a non-existing result spawns lots of question regarding duration under consideration (how long to evaluate?), appropriateness of technology (is all the factors of the requirement is captured by selected technologies?), and so on.

Similarly, in services like online shopping, the physical existence cannot be a criterion. In online shopping, we browse, select and buy the items from a buyer mostly using a payment gateway provided by another vendor. Although the payment gateway is a part of online shopping service machine, the payment gateway is a service by itself. The question would be should we consider this shopping and payment as single service machine, or as two service machine, or one service machine as a subsystem of another service machine.

The introduction of wearable technology, Internet of Everything (IoE), and Artificial Intelligence (AI) into the services would impact the way services are rendered. With the data-driven approach which could capture the variability in customer demand and behavior, the services become more and more customizable, and the providers would be able to provide personalized services. Many services could be configured as self-service or service-factory. Although the service quality

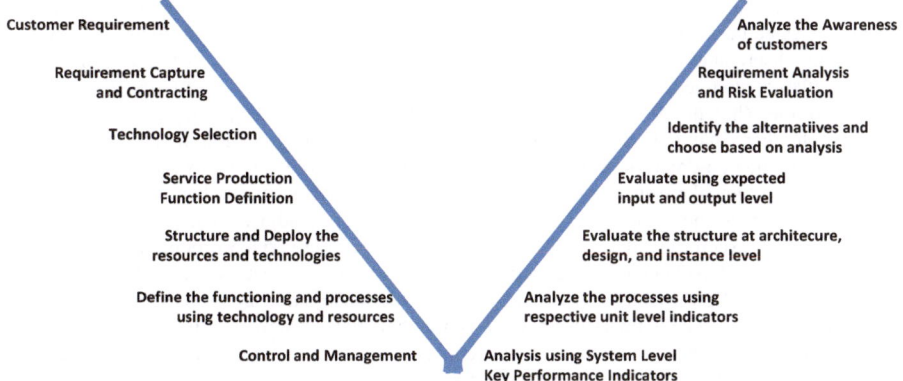

Fig. 5.1 Analysis of service machine

improves with these advancements, they come with a cost of privacy and safety of the customer.

> Google as a service provides the most relevant search results for you, but it captures the data regarding your behavior and interests in possible ways. At the same time, another search service engine DuckDuckgo provides less relevant search results many times without storing your data. It is crucial for the customer to make decisions about the trade-offs to be made. The similar instances could be noticed in different services such as Facebook, Uber, and so on. As a service machine, the customer needs to understand its working, pros and cons and possible risks clearly to make appropriate trade-off decisions.

The ethical and legal aspects of the service machine are significant in the current context. The service providers may need lots of data to make their services useful. But it should not be violating the norms of the nation in which it operates. The surveillance nature of the services is not acceptable for most of the people in a country. At the same time, people would like to share the day-to-day developments in their personal and professional fronts. As a service provider, there is a need to understand the local and global legal, ethical and social aspects to provide services. In the medical sector, the pharmaceuticals and hospitals would have similar problems in designing service machine for clinical trials.

The expected use of Service Machine concept is to diagnose the issues in a service production system, similar to diagnosing problems in the physical machine. Usually, in physical machines, we deploy scientific/engineering methods for diagnosis/prognosis. We debug the system and isolate the problem area and solve it. Probably, we can use the "V" model usually used in system testing for diagnosing the service machine as shown in Fig. 5.1.

> In EMS if we noticed that the service machine fails to save and stabilize the expected number of patients, we have to investigate the working of EMS service machine. Many times it happens that we control and manage upon a wrong KPI and focus of operations become wrong. Although the fast service increases the chances of survival of the patient, if response time or service time is the only performance criteria for the functioning, the actors involved may not focus on saving or stabilizing the patient instead look for providing quick services.

If we are capturing right performance indicators and noticed the indicators are not reaching the expected performance, we need to look into functioning and processes involved. If we find any discrepancy from the expected operation, it can be rectified. We may notice that all the calls are attended, and ambulances are dispatched to the patients within the stipulated time frame. Thus the problem may occur in structure. The architecture may be flawed, for example, there may not have enough communication or communication medium between different value co-creators. Similarly, the design can be flawed – We may have selected a larger geographical area to serve, or we may allocate an insufficient number of ambulances. It may be due to the not enough flexibility in the system to readjust it with the demand. If we see that structurally the service should work as per the service production function defined we need to check if service production functions are correctly defined. The evaluation may show that the technologies used in the production function are flawed. Furthermore, there is a chance that the interfaces not good enough to make customers utilize the technologies. Sometimes we may identify that the people are not aware of their emergency need and report same lately which in turn results in bad performance of the system.

Sometimes, the SM fails to cater the demand due to the mismatch between the sub-components of the machine template. It may eventually lead to the failure of SM.

The increase in the market share of Gmail compares to Yahoo! Mail attribute to the mismatch between the service production function and the structure. One may able to see that service production function of both the service machine are same, and they may match in their architecture and design. At some point of time, Gmail as a service machine provided more flexibility to the customer in terms integrating different items such as search, instant messaging, and contacts to a single umbrella, spam control, and simple, non-cluttered interface. But, both of the services work with the principle of getting revenue from the advertisers for free service to email users. In the recent scenario, there are many competitive services are emerging as these services compromised the customer's privacy. Thus, there are some email service providers like Protonmail (from Proton Technologies AG) introduced the service production function to ensure the privacy and data security. The complete services are available at a small premium.

With the lens provided by SM template may able to see the missing links. The failure of some SM attributes to the lack of appropriate technologies. The prevention and care services are mostly self-service SM. It follows that the customer needs to participate in value creation to a higher degree. But due to unavailability of appropriate behavioral technologies to make customer contribute to the required level results in many failures.

The patient's adherence to the prescription and medical advice post-treatment is essential for the value creation. Due to the multiplicative nature of value in services, the non-adherence may lead to value deduction. The behavioral technologies are crucial to monitor and motivate the patients to adhere to the advice. SM for de-addiction, preventive care, and so on fails due to lack of appropriate technology. In the world of online services, there are different services overcrowded with spammed and fake information and users. The lack of technologies both information and behavioral makes the providers challenging to moderate the operations appropriately.

The service machine needs a dynamic upgrade with the available technologies, else the machine may become obsolete.

The online social network Facebook overtook another social network due to the weak implementation of the behavioral technologies to engage people. Facebook introduces one or other behavioral technology to "hook" the users to it time to time. Orkut apparently failed to do so and lost in the competition with Facebook.

Thus the failure of SM is primarily due to missing links (failed interfaces), inappropriate definition or use of components (flawed component), and most importantly engaging the customers. By evaluating a service machine using Service Machine Template, the service entrepreneurs/ researchers can identify the loopholes and fix them. The advancements in the fields of (big) data analytics, process mining, communication, and artificial intelligence would help the service machine designers in understanding and prioritize the requirements and align the service production functions accordingly.

References

Al-Shaqsi, S. Z. K. (2010). Response time as a sole performance indicator in EMS: Pitfalls and solutions. *Open Access Emergency Medicine, 2*, 1–6.

Baek, J. S., Kim, S., Pahk, Y., & Manzini, E. (2017). A sociotechnical framework for the design of collaborative services. *Design Studies, 55*, 54–78.

Baumol, W. J. (1967). Macroeconomics of unbalanced growth: The anatomy of urban crisis. *The American Economic Review, 57*(3), 415–426.

Christensen, C. M., Hall, T., Dillon, K., & Duncan, D. S. (2016). *Competing against luck: The story of innovation and customer choice* (Latest edition). New York: Harper Business.

Fitzsimmons, J. A., & Fitzsimmons, M. J. (2010). *Service management: Operations, strategy, information technology* (7th Rev. ed.). New York: McGraw-Hill.

Fry, M., & Burr, G. (2002). Review of the triage literature: Past, present, future? *Australian Emergency Nursing Journal, 5*(2), 33–38. https://doi.org/10.1016/S1328-2743(02)80018-9

Fry, M., & Stainton, C. (2005). An educational framework for triage nursing based on gatekeeping, timekeeping and decision-making processes. *Accident and Emergency Nursing, 13*(4), 214–219. https://doi.org/10.1016/j.aaen.2005.09.004

Glaser, B., & Strauss, A. (1999). *The discovery of grounded theory: Strategies for qualitative research*. Chicago: Aldine Transaction.

Hill, T. P. (1977). On goods and services. *Review of Income and Wealth, 23*(4), 315–338.

Hopp, W. J., & Spearman, M. L. (2011). *Factory physics* (3rd ed.). Long Grove, IL: Waveland Pr Inc.

Lidal, I. B., Holte, H. H., & Vist, G. E. (2013). Triage systems for pre-hospital emergency medical services – A systematic review. *Scandinavian Journal of Trauma, Resuscitation and Emergency Medicine, 21*, 28–28. https://doi.org/10.1186/1757-7241-21-28

Lillrank, P. (2018). *The logics of healthcare: The professional's guide to health systems science* (1st ed.). Milton, UK: Productivity Press.

Lillrank, P., Groop, P. J., & Malmström, T. J. (2010). Demand and supply–based operating modes – A framework for analyzing health care service production. *Milbank Quarterly, 88*(4), 595–615.

Loudon, K. C., & Loudon, J. P. (2015). *Management information systems: Managing the digital firm* (14th ed.). Boston: Pearson.

Maglio, P. P., & Spohrer, J. (2008). Fundamentals of service science. *Journal of the Academy of Marketing Science, 36*(1), 18–20. https://doi.org/10.1007/s11747-007-0058-9

McGowan, M. A., Andrews, D., & Millot, V. (2017). *The walking dead? Zombie firms and productivity performance in OECD countries*. OECD Economics Department Working Paper, No. 1372.

Moeller, S. (2010). Characteristics of services – A new approach uncovers their value. *Journal of Services Marketing, 24*(5), 359–368. https://doi.org/10.1108/08876041011060468

National Highway Traffic Safety Administration. (1995). *"Star of life", Emergency medical care symbol: Background, specifications, and criteria*. U.S. Department of Transportation, National Highway Traffic Safety Administration, Office of Enforcement and Emergency Services. Retrieved from http://www.ems.gov/vgn-ext-templating/ems/sol/index.htm

National Highway Traffic Safety Administration (DOT), W., DC, & Health Resources and Services Administration (DHHS/PHS), W., DC. Maternal and Child Health Bureau. (1996). *Emergency medical dispatch. National standard curriculum. Instructor guide. Trainee guide*. U.S. Government Printing Office, Superintendent of Documents, Mail Stop: SSOP, Washington, DC 20402-9328.

Osterwalder, A., & others. (2004). *The business model ontology: A proposition in a design science approach*. Doctoral dissertation. Lausanne, Switzerland: University of Lausanne.

Osterwalder, A., & Pigneur, Y. (2010). *Business model generation: A handbook for visionaries, game changers, and challengers*. Hoboken, NJ: Wiley.

Robers, C. (2011). *Demand increases for air ambulance use in Texas*. Bound Tree University. Retrieved from http://www.boundtreeuniversity.com/Specialized-Rescue/news/1046709-Demand-increases-for-air-ambulance-use-in-Texas

Sampson, S. E., & Froehle, C. M. (2006). Foundations and implications of a proposed unified services theory. *Production and Operations Management, 15*(2), 329–343. https://doi.org/10.1111/j.1937-5956.2006.tb00248.x

Sandori, P. (1982). *The logic of machines and structures*. Mineola, NY: Courier Dover Publications.

Schmenner, R. W. (2004). Service businesses and productivity. *Decision Sciences, 35*(3), 333–347.

Spaite, D. W., Criss, E. A., Valenzuela, T. D., & Guisto, J. (1995). Emergency medical service systems research: Problems of the past, challenges of the future. *Annals of Emergency Medicine, 26*(2), 146–152. https://doi.org/10.1016/S0196-0644(95)70144-3

Vargo, S. L., & Lusch, R. F. (2004). Evolving to a new dominant logic for marketing. *Journal of Marketing, 68*(1), 1–17. https://doi.org/10.1509/jmkg.68.1.1.24036

Weick, K. E. (1995). *Sensemaking in organizations* (Vol. 3). Thousand Oaks, CA: Sage.

Wemmerlöv, U. (1990). A taxonomy for service processes and its implications for system design. *International Journal of Service Industry Management, 1*(3), 20–40.

Womack, J. P., Jones, D. T., & Roos, D. (1990). *Machine that changed the world*. London: Simon and Schuster.

Zeithaml, V., Bitner, M. J., & Gremler, D. (2012). *Services marketing* (6th ed.). New York: McGraw-Hill Education.